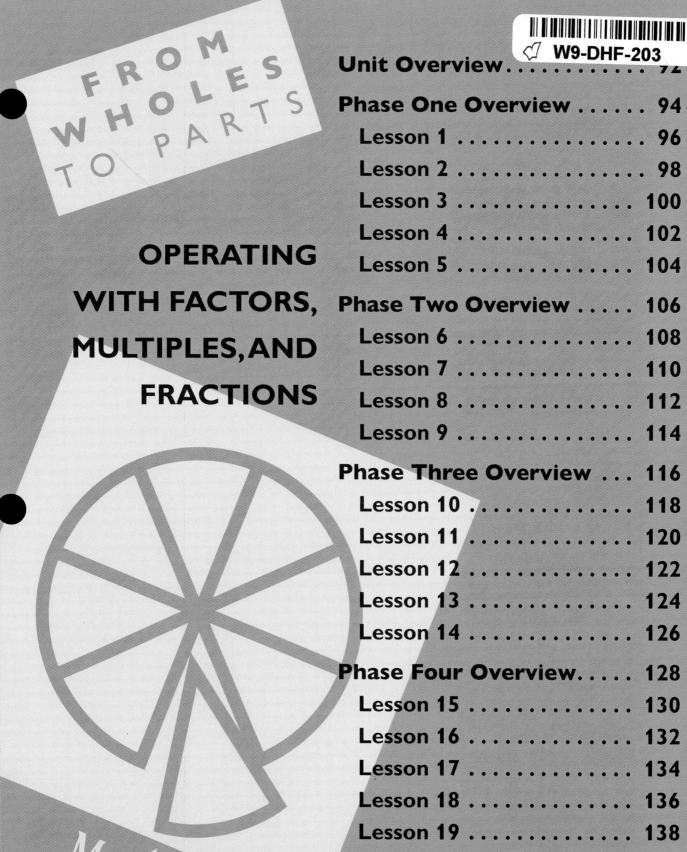

FROM WHOLES TO PARTS

OPERATING WITH FACTORS, MULTIPLES, AND FRACTIONS

MathScape
SEEING AND THINKING MATHEMATICALLY

PHASE**ONE**
The Whole of It

To work with fractions, it helps to be at ease with whole numbers. In this phase, you will focus on whole numbers. As you build models, figure out computational methods, and play fast-paced games, you will make connections among factors, multiples, and prime numbers.

How can you compute with numbers that are not whole numbers?

FROM WHOLES TO PARTS

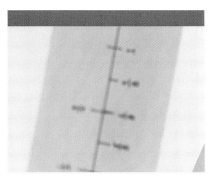

PHASE**TWO**
Between the Whole Numbers

Now it is time to focus on the parts of the whole. Making designs will help you to see how fractional parts relate to each other and the whole. By using number lines, you will discover ways to compare fractions. Games with cards and number cubes will make it lots of fun to try your new skills.

PHASE**THREE**
Adding Parts and Taking Them Away

With all that you have learned, you will be a natural at adding and subtracting fractions. You will have the chance to use area models, number lines, and computation. At the end of the phase, you will help a defective robot get out of a room and teach an extra-terrestrial visitor how to add and subtract fractions.

PHASE**FOUR**
Fractions in Groups

When you multiply and divide fractions, the answers are easy to compute. But, what is really happening? You will model fractions to sort out the confusing truth – sometimes multiplying results in a lesser answer and sometimes dividing results in a greater answer. How does this happen? You will figure out the hows and whys in this phase.

PHASE ONE

You will not see any fractions in this phase, but you will learn things that make working with fractions easier. Modeling shapes will help you find prime factors, which you can use to find greatest common factors. You will also find a way to identify least common multiples. You will look at the rules of order for solving problems and discover ways you can and cannot rearrange numbers in a problem to make it easier to solve.

The Whole of It

WHAT'S THE MATH?

Investigations in this section focus on:

NUMBER and OPERATIONS

- Understanding factors and prime factors
- Finding factors of whole numbers
- Finding the prime factorization of a whole number
- Finding the greatest common factor of two or more whole numbers
- Generating multiples of a number
- Finding the least common multiple of two or more numbers
- Performing operations in the correct order

MathScape Online
mathscape1.com/self_check_quiz

Shapes and Factors

MODELING
NUMBERS TO
IDENTIFY FACTORS

What can rectangles tell you about factors? To start this phase about whole numbers, you will do some model building. Then, you will take your modeling to a third dimension using cubes.

Find Factors Using Rectangles

What are the possible side lengths of a rectangle made from a specific number of unit squares?

Elaine uses square tiles to build tables with mosaic tops. Your class is going to find all the rectangular tabletops Elaine can make using specific numbers of tiles. For each number of tiles, you will make solid rectangles, in one layer, using all of the tiles.

Experiment until you are sure you have all the possible rectangles for each number you are assigned. Then, write the side lengths for each rectangle on the class chart.

1 Which numbers of tiles can produce more than one rectangle?

2 Which numbers of tiles can produce only one rectangle?

Dimensions and Factors

You can count the number of tiles on each side of a rectangle to find the lengths or **dimensions** of the sides.

Each dimension is a **factor** of the total number of tiles in the rectangle. A factor is a number that is multiplied by another number to yield a product. For example, 8 and 2 are factors of 16.

96   FROM WHOLES TO PARTS • LESSON 1

Find Factors Using Cubes

Since rectangles have two dimensions, each tile tabletop you modeled showed you two factors. Now, you will use cubes to build three-dimensional shapes called rectangular prisms.

What can three-dimensional modeling tell you about factors?

1 With your group, try to determine what numbers from 1 to 30 can be modeled as rectangular prisms without using 1 as a dimension. When you find a number for which you can make such a prism, see whether you can make a different prism with the same number of cubes. Your group should make a chart that shows:

- the total number of cubes, and

- the dimensions of each prism you made.

The number 8 can be modeled as a rectangular prism without using 1 as a dimension.

2 Compare your group's chart to the chart the class made for tile tabletops. How can you use the tabletop chart to predict the prism chart?

3 For which number could you make the most prisms? How many different prisms could you make for that number? What allows that number to be modeled in so many ways?

4 What factors do the following numbers have in common?

a. 8 and 15

b. 9 and 24

c. 12 and 16

d. 15 and 18

e. 20 and 24

f. 24 and 30

Write a Definition

On your copy of My Math Dictionary, write your own definitions and give examples of the following terms:

- factor

- prime factor

- common factor

prime number
factor
common factor

page 142

2 The Great Factor Hunt

As you saw when you built rectangles and prisms, numbers can be written as products of two, three, or more factors. How do you know when you have found the most possible factors for a number?

Find All the Factors

What is the longest string of factors you can multiply to get a given number?

One way to express the number 12 is to show it as a product of two factors: $12 = 2 \times 6$. How would you express the number 12 as the product of more than two factors? Without using 1 as a factor, what is the longest string of factors you can use to express the number 12?

Product of two factors: $12 = 2 \times 6$

Longest string of factors: $12 = 2 \times 2 \times 3$

The longest string of factors, excluding the factor 1, is called the *prime factorization* of the number.

Work with your classmates to find the longest factor string for each whole number from 1 to 30. When the class chart is finished, answer the following questions.

1 What kind of numbers cannot be part of these factor strings? What kind of numbers can be part of these factor strings?

2 Find three numbers that have a prime factor in common. List the numbers and the factor they share.

3 Find two pairs of numbers that have a string of more than one factor in common. List the number pairs and the factors they share.

$16 = 2 \times 8$ $18 = 2 \times 9$ $27 = 3 \times 9$

$\quad = 2 \times 2 \times 4$ $= 2 \times 3 \times 3$ $= 3 \times 3 \times 3$

$\quad = 2 \times 2 \times 2 \times 2$

Find the Greatest Common Factor

Look at the prime factorization chart that your class compiled. Compare the prime factor strings for 12 and 30.

$$12 = 2 \times 2 \times 3$$

$$30 = 2 \times 3 \times 5$$

The numbers 12 and 30 have 2×3 in common. We will call "2×3" a "common string."

1 Write the longest "common string" of factors for each pair of numbers.

a. 9 and 27   **b.** 8 and 24

c. 24 and 30   **d.** 45 and 60

2 Use each pair's "common string" to find the greatest factor they share. For example, the longest "common string" of 12 and 30 is 2×3. The greatest factor they share is 2×3 or 6. This number is called the *greatest common factor* of 12 and 30.

3 Write each number in each pair as a product of their greatest common factor and another factor. For example, $12 = 6 \times 2$ and $30 = 6 \times 5$.

$12 = 2 \times \boxed{2 \times 3}$

$30 = \boxed{2 \times 3} \times 5$

$2 \times 3 = 6$ 6 is the greatest common factor of 12 and 30.

$12 = 6 \times 2$

$30 = 6 \times 5$

What is a quick way to identify a greatest common factor?

Write a Definition

On your copy of My Math Dictionary, write your own definitions and give examples of the following terms:

- prime factorization

- greatest common factor

hot **words** | prime factorization
greatest common factor

Homework

page 143

3 Multiple Approaches

When working with fractions, you often need to use common multiples of two or more numbers. In this lesson, you will use playing cards to find common multiples. Then, you will learn to find the least common multiple of any pair of numbers.

Play "It's in the Cards"

How can you use playing cards to determine common multiples?

You can use a standard deck of cards to study common multiples.

1 With a partner, try the example game of "It's in the Cards" shown below. What common multiples do the cards show?

2 Play the game three more times using the rates 3 and 4, 2 and 5, and 3 and 6. Make a table showing each pair of numbers and the multiples you find for each pair.

3 For each number pair, circle the least multiple that you found.

4 Using your own words, define the terms *common multiple* and *least common multiple*.

It's in the Cards

- Each player has a stack of 13 playing cards, representing the numbers 1 (ace) to 13 (king), in numerical order.

- Each player lays each of his or her cards facedown in a line.

- Each player turns over certain cards according to his or her rate. In the example, Player A had a rate of 2, so every second card is turned. Player B had a rate of 3, so every third card is turned.

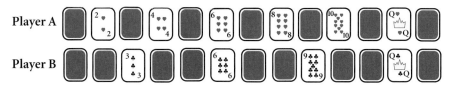

The numbers that are faceup in both rows are common multiples of the rate numbers.

Find the Least Common Multiple

You may have noticed that the product of any two numbers is always a multiple of both numbers. For example, 54 is a common multiple of 6 and 9. But how can you find the least common multiple?

How can prime factors help you find least common multiples?

1 The table below shows steps for finding the least common multiple of various number pairs. Try to determine what happens in each step. Consider these questions.

- What prime factors do the given numbers have in common?

- How does the number in step 2 relate to the given numbers?

- Each given number is written as a product in step 3. What factors are used?

- Which numbers from step 3 are used in step 4?

2 When you think you know a method for finding the least common multiple of two numbers, use the method to find the least common multiple of 12 and 18, 14 and 49, and several other pairs of numbers. Does your method work? Make a table showing each pair of numbers, the steps you used, and the least common multiple of the numbers.

3 Describe your method for finding the least common multiple of two numbers.

4 How can you use your method to find the least common multiple of three numbers?

Steps to Find the Least Common Multiple

Given Numbers	Step 1	Step 2	Step 3	Step 4	Least Common Multiple
8	$2 \times 2 \times 2$	4	$8 = 4 \times 2$	$4 \times 2 \times 3$	24
12	$2 \times 2 \times 3$		$12 = 4 \times 3$		
15	3×5	5	$15 = 5 \times 3$	$5 \times 3 \times 14$	210
70	$2 \times 5 \times 7$		$70 = 5 \times 14$		
30	$2 \times 3 \times 5$	15	$30 = 15 \times 2$	$15 \times 2 \times 3$	90
45	$3 \times 3 \times 5$		$45 = 15 \times 3$		

hot **words** | multiple
least common multiple

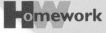

Homework

page 144

4 First Things First

How much can you do with 1, 2, and 3? If you know the rules about the order of operations, you can make three numbers do a lot. In this lesson, you will use the rules to expand the possibilities.

Use the Order of Operations

In what order should an equation's operations be computed?

Work on your own to solve each of the following equations.

$$1 + 3 \times 2 = n \qquad 3 + 1 \times 2 = n \qquad 4 + 6 \div 2 = n$$

What answers did you get? Did everyone in your class get the same answers? Although everyone's computations may appear to be correct, why might some people get a different answer than you did?

Each of these problems has just one correct answer. To get the correct answer, you must follow the order of operations.

Your teacher will give you a handout. See if you can place the numbers 1, 2, and 3 in all ten equations to get a result of every whole number from 1 to 10. Make sure you follow the order of operations.

Order of Operations

- First, do any computations that are within **grouping symbols** such as parentheses.
- Then, evaluate any **exponents.**
- Next, **multiply** and/or **divide** in order from left to right.
- Finally, **add** and/or **subtract** in order from left to right.

Play "Hit the Target"

In this game, you will make equations using three numbers to equal a target number. Your team will have a better chance of winning if you know the order of operations.

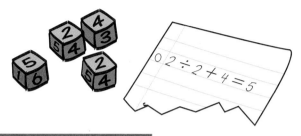

Can you get the target value using the order of operations?

Hit the Target

For each round of the game, your teacher will roll four number cubes. One is the target number. The other three are the building numbers. Your goal is to use all three building numbers and two operations to get as close to the target number as possible. Building numbers may be used in any order, but each number may only be used once.

Each team earns points as follows.

- One point for getting the target number.

- An additional point if you get the target number without using parentheses.

- An additional point if you get the target number using a combination of addition or subtraction and multiplication or division.

All teams that get the target number earn points. If no team gets the target number, the team with the number closest to the target number earns a point.

Write a Memory Helper

As you played "Hit the Target," you had to think fast and keep the order of operations in mind. What was your method of remembering the order of operations? Can you think of something that would help you remember? Write down your method and share it with the class.

hot **words** | operations
order of operations

Homework
page 145

5 Putting It All Together

SOLVING PROBLEMS
WITH WHOLE
NUMBERS

Can you change the order of numbers or the parentheses in an expression without changing its value? Sometimes you can and sometimes you cannot. In this lesson, you will learn when you can change them. Then, you will solve "guess my number" problems.

Organize Your Math

What changes can you make in an expression to make it easier to find its value?

Look at the following pairs of expressions carefully. For each pair, try to predict whether the value of **A** will be the same as the value of **B**. If expression **A** can be changed to expression **B** without changing the value of **A**, explain why. If expression **A** cannot be changed to expression **B** without changing the value of **A**, explain why not.

1 A $3 + 34 + 27$

  B $3 + 27 + 34$

3 A $(27 - 7) - 3$

  B $27 - (7 - 3)$

5 A $(24 \div 4) \div 2$

  B $24 \div (4 \div 2)$

7 A $24 \div 8 + 2$

  B $24 \div 2 + 8$

9 A $5 + 5 \times 14$

  B $(5 + 5) \times 14$

2 A $524 - 412$

  B $412 - 524$

4 A $4 \times (8 \times 2)$

  B $(4 \times 8) \times 2$

6 A 18×6

  B $10 \times 6 + 8 \times 6$

8 A 208×4

  B $(200 \times 4) + (8 \times 4)$

10 A $4 \times 6 + 5$

  B $6 \times 4 + 5$

11 Write an expression of your own that could be made easier by a change in order or grouping. Show what change you would make. Then, describe the change. Explain:

- why it makes finding the value easier, and

- why it does not change the value.

Guess My Number

Now, you will use what you have learned about factors and multiples to find mystery numbers.

Can you find the mystery numbers?

1. Two numbers have a sum of 60. Both are multiples of 12. Neither number is greater than 40. What are the numbers?

2. Three numbers are each less than 20. They are all odd. One is the least common multiple of the other two numbers. What are the numbers?

3. The greatest common factor of two numbers is 11. One number is twice the other. Their least common multiple is one of the numbers. What are the numbers?

4. Sam and Susanna are brother and sister. The difference between their ages is a factor of each of their ages. Their combined age is 15. What are their possible ages?

5. Luis is 6 years older than his dog. His age and his dog's age are both factors of 24, but neither is a prime factor. What are their ages?

6. The number of pennies saved by Brian is a multiple of the number of pennies saved by his sister Carmen. Together, they have saved 125 pennies. Carmen has saved more than 10 pennies. Both numbers of pennies are multiples of 5. How many pennies does each have?

Write Your Own Problem

Create a problem of your own that involves factors or multiples. When you have finished, give your problem to another student and see if she or he can guess your numbers.

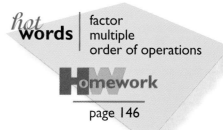

hot **words** | factor
multiple
order of operations

Homework

page 146

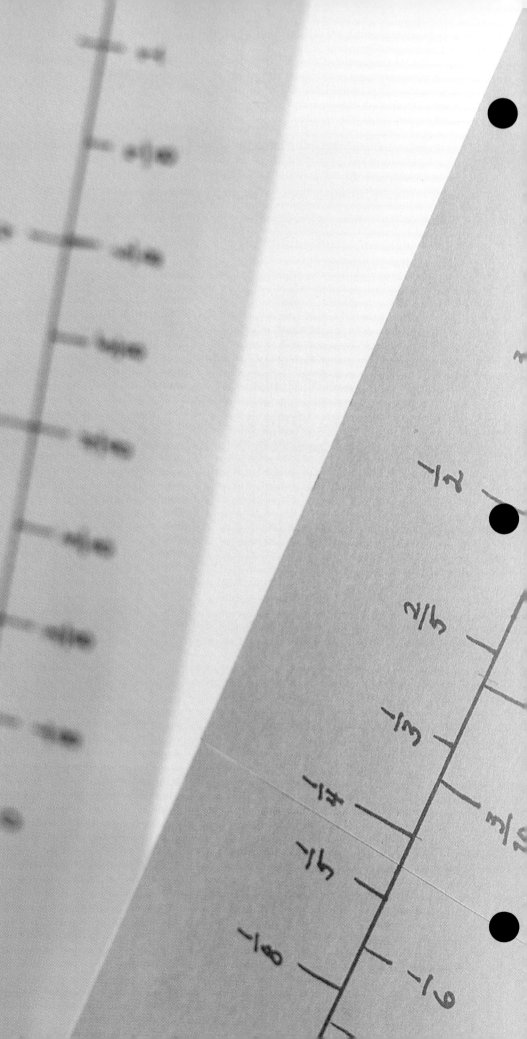

PHASE TWO

In daily life, you deal with fractions – a quarter pound, two and a half hours, six tenths of a mile – without thinking much about it. The experiences you have in this phase will give you a fuller understanding of fractions. You will make area models to show how different parts of a whole relate to each other. You will use number lines not only to order and compare fractions, but to see that every fraction has many names.

Between the Whole Numbers

WHAT'S THE MATH?

Investigations in this section focus on:

NUMBER and OPERATIONS

- Representing fractions as parts of a whole
- Writing an equivalent fraction for a given fraction
- Comparing and ordering fractions
- Graphing fractions on a number line
- Understanding improper fractions
- Finding a common denominator for two fractions

MathScape Online
mathscape1.com/self_check_quiz

Designer Fractions

DIVIDING WHOLE
AREAS INTO
FRACTIONAL PARTS

Can you be a math-savvy designer? In this lesson, you will divide whole rectangles into equal parts. Then, you will make designs with parts that are not the same size, but still represent one whole.

Make Designs to Show Fractions

How many different ways can you divide a rectangle into four equal parts?

Fractions can be represented as parts of a rectangle.

1 On centimeter grid paper, make a rectangle that is 8 centimeters by 10 centimeters and divide it into four equal parts. Each person in your group should make a different design using the following guidelines.

- The parts may look exactly alike or they may be different.

- You may divide parts with diagonal lines.

When you have finished, discuss with your group how you know that the four parts in your design are equal.

2 On centimeter grid paper, make four rectangles that are 6 centimeters by 8 centimeters. Your group will be assigned one of the following pairs of fractions.

$$\frac{1}{3} \text{ and } \frac{1}{6} \qquad \frac{1}{6} \text{ and } \frac{1}{12} \qquad \frac{1}{3} \text{ and } \frac{1}{12}$$

Your group should make two different designs for each of your two fractions. Then, answer these questions about your designs.

a. How do you know which of the fractional parts in the different designs are equal to each other?

b. Compare the sizes of the fractional parts for each of your two fractions. What do you notice?

c. What other fractions would be easy to make using the same rectangle?

Use Fractions to Make a Whole

You can make designs where each region represents a given fraction.

1 Make one design for each group of fractions. First, outline the rectangle on grid paper using the dimensions given. Then, divide the rectangle into the fractional parts. Color each part and label it with the correct fraction.

a. On a 2-by-4-centimeter rectangle, show $\frac{1}{2}$, $\frac{1}{4}$, $\frac{1}{8}$, and $\frac{1}{8}$.

b. On a 3-by-4-centimeter rectangle, show $\frac{1}{3}$, $\frac{1}{3}$, $\frac{1}{6}$, and $\frac{1}{6}$.

c. On a 3-by-4-centimeter rectangle, show $\frac{1}{4}$, $\frac{1}{4}$, $\frac{1}{6}$, $\frac{1}{6}$, and $\frac{1}{6}$.

d. On a 4-by-6-centimeter rectangle, show $\frac{1}{4}$, $\frac{1}{8}$, $\frac{1}{8}$, $\frac{1}{3}$, and $\frac{1}{6}$.

2 Make a design to help determine the missing fraction in each statement. Label each part with the correct fraction.

a. Use a 3-by-4-centimeter rectangle to show
$$\frac{2}{3} + \frac{1}{4} + \underline{\ ?\ } = 1.$$

b. Use a 3-by-4-centimeter rectangle to show
$$\frac{1}{3} + \frac{1}{4} + \frac{1}{4} + \underline{\ ?\ } = 1.$$

> **What designs can you make with different fractions that together make a whole?**

Make Your Own Design

Create a design of your own that shows several fractions whose sum is one. Use the following guidelines.

- Use a different group of fractions than those listed above. Think about what size of rectangle would make sense for the group of fractions you choose.

- Divide your rectangle into at least four parts.

- Use at least four different fractions in your design.

- Label each part with the correct fraction.

hot **words** | fraction

Homework

page 147

 **Area Models and Equivalent Fractions**

How can you compare fractions? In this lesson, you will use area models and grid sketches.

Use Area Models to Compare Fractions

Which fraction is greater?

It's easy to compare two fractions like $\frac{1}{3}$ and $\frac{2}{3}$ because they have the same denominators. But what about fractions with different denominators? You can compare these fractions by using area models.

Use area models to compare each set of fractions.

1 $\frac{4}{5}$ and $\frac{2}{3}$ **2** $\frac{2}{7}$ and $\frac{1}{3}$ **3** $\frac{4}{5}$ and $\frac{3}{4}$

Using Area Models to Compare Fractions

To compare $\frac{1}{3}$ and $\frac{2}{5}$, make two 5-by-3 rectangles on grid paper. To show $\frac{1}{3}$, circle one of the columns. To show $\frac{2}{5}$, circle two rows.

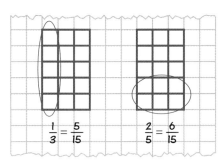

$$\frac{1}{3} = \frac{5}{15} \qquad \frac{2}{5} = \frac{6}{15}$$

Since 5 of the 15 squares cover the same area as $\frac{1}{3}$, $\frac{5}{15}$ is equal to $\frac{1}{3}$. So $\frac{5}{15}$ and $\frac{1}{3}$ are **equivalent fractions.** Also, $\frac{6}{15}$ is **equivalent** to $\frac{2}{5}$. It is easier to compare $\frac{5}{15}$ and $\frac{6}{15}$ than it is to compare $\frac{1}{3}$ and $\frac{2}{5}$. So, $\frac{1}{3} < \frac{2}{5}$.

Use Grid Sketches to Compare Fractions

A quick sketch on one grid can be helpful for comparing fractions.

This sketch shows the comparison between $\frac{2}{5}$ and $\frac{1}{3}$.

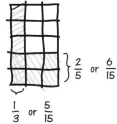

$\frac{6}{15}$ is greater than $\frac{5}{15}$.

How can a grid sketch help you to compare fractions?

1 Make a grid sketch for each pair of fractions. Label each sketch with the given fractions and their equivalent fractions. Compare the fractions.

a. $\frac{2}{3}$ and $\frac{3}{5}$ **b.** $\frac{2}{8}$ and $\frac{1}{5}$

c. $\frac{1}{3}$ and $\frac{2}{6}$ **d.** $\frac{6}{7}$ and $\frac{4}{6}$

2 Write a few sentences explaining how a sketch helps you decide which fraction is greater.

Practice Fraction Comparisons

Two fractions with the same denominator have a *common* denominator.

How can you use common denominators to compare fractions?

1 For each pair of fractions, find equivalent fractions that have a common denominator. Determine which fraction is greater. Write your answer using the original fractions.

a. $\frac{5}{6}$ and $\frac{7}{10}$ **b.** $\frac{4}{6}$ and $\frac{3}{4}$

c. $\frac{7}{8}$ and $\frac{4}{5}$ **d.** $\frac{3}{4}$ and $\frac{7}{10}$

2 Describe how you prefer to compare fractions. Why does your method work? If you prefer one method for some fractions and a different method for others, explain.

3 In your copy of My Math Dictionary, write a definition of *equivalent fractions*. Use a drawing to illustrate your definition.

| equivalent fractions |
| denominator |

page 148

8 Fraction Lineup

PLOTTING NUMBERS
ON THE NUMBER
LINE

What numbers belong between the whole numbers on a number line? In this lesson, you will be placing fractions in appropriate places on a number line. You will start with separate lines for different fractions and then combine them all onto one number line.

Place Fractions on a Number Line

What are the positions of fractions on a number line?

The number line below shows only whole numbers.

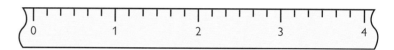

You can place fractions between the whole numbers.

1 Mark fractions on separate number lines as indicated.

  a. On number line **A**, mark all the halves.

  b. On number line **B**, mark all the thirds.

  c. On number line **C**, mark all the fourths.

  d. On number line **D**, mark all the sixths.

  e. On number line **E**, write all the fractions that you wrote on lines **A** through **D**. Where two or more fractions have the same position, write each fraction.

2 Use your number lines to compare each pair of fractions. Write an expression using $<$, $>$, or $=$ to show the comparison.

  a. $\frac{1}{4}$ and $\frac{2}{6}$ **b.** $\frac{4}{3}$ and $\frac{8}{6}$ **c.** $\frac{5}{3}$ and $\frac{10}{6}$

  d. $\frac{5}{6}$ and $\frac{3}{4}$ **e.** $\frac{10}{6}$ and $\frac{7}{4}$ **f.** $\frac{3}{2}$ and $\frac{8}{6}$

3 Explain how you can use a number line to compare two fractions.

Use the Number Line to Find Equivalent Fractions

In Lesson 7, you learned that equivalent fractions cover the same area on an area model. Equivalent fractions are also located at the same place on a number line.

How can you use a number line to identify equivalent fractions?

1 Find an equivalent fraction for each fraction.

    a. $\dfrac{3}{2}$ **b.** $\dfrac{5}{3}$ **c.** $\dfrac{6}{3}$

    d. $\dfrac{4}{6}$ **e.** $\dfrac{6}{6}$ **f.** $\dfrac{4}{3}$

2 Imagine that you were to label all the twelfths $\left(\dfrac{0}{12}, \dfrac{1}{12}, \dfrac{2}{12}, \text{and so on}\right)$ on your number line **E**. Which fraction(s) that you had already labeled are equivalent to $\dfrac{3}{12}$? to $\dfrac{10}{12}$? to $\dfrac{18}{12}$?

3 The number line below is marked in ninths. Which of these fractions would be equivalent to fractions that you labeled on number line **E**?

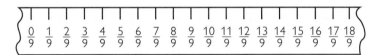

$$\frac{0}{9} \quad \frac{1}{9} \quad \frac{2}{9} \quad \frac{3}{9} \quad \frac{4}{9} \quad \frac{5}{9} \quad \frac{6}{9} \quad \frac{7}{9} \quad \frac{8}{9} \quad \frac{9}{9} \quad \frac{10}{9} \quad \frac{11}{9} \quad \frac{12}{9} \quad \frac{13}{9} \quad \frac{14}{9} \quad \frac{15}{9} \quad \frac{16}{9} \quad \frac{17}{9} \quad \frac{18}{9}$$

Write about Equivalent Fractions

Write two fractions that have a common denominator. Make one equivalent to $\frac{3}{4}$ and the other equivalent to $\frac{5}{6}$. Explain why they are equivalent. Include both an area model and a number line with your answer.

hot **words** | equivalent fractions | improper fraction

Homework

page 149

9 Focus on Denominators

Writing equivalent fractions with common denominators makes them easier to compare. However, when you use the product of the denominators as the common denominator, this number may be larger than you need.

How can you be sure that you have found the least common denominator?

Look for the Least Common Denominator

You studied least common multiples in Lesson 3. You can use least common multiples to find least common denominators.

1 For each pair of fractions, complete the following:

- Find the product of the denominators.

- Find the least common multiple of the denominators.

- Write equivalent fractions with common denominators using the lesser of the two numbers, if they are different.

- Compare the fractions using the symbols $<$, $>$, or $=$.

  a. $\frac{5}{6}$ and $\frac{7}{12}$ **b.** $\frac{3}{10}$ and $\frac{1}{6}$ **c.** $\frac{2}{3}$ and $\frac{3}{4}$

  d. $\frac{4}{5}$ and $\frac{7}{10}$ **e.** $\frac{2}{7}$ and $\frac{1}{4}$ **f.** $\frac{2}{3}$ and $\frac{7}{9}$

2 In step **1**, which method of finding a common denominator gave the lesser common denominator for the fraction pairs?

3 Using the number line below, find equivalent fractions with the least common denominator for $\frac{5}{10}$ and $\frac{4}{6}$.

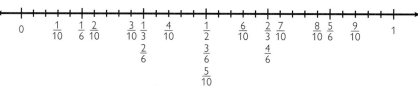

4 Explain why the least common multiple of the denominators of $\frac{5}{10}$ and $\frac{4}{6}$ is not their least common denominator. What should you do before finding the least common denominator?

Play "Get the Cards in Order"

Play "Get the Cards in Order" with a partner.

Can you order fractions?

Get the Cards in Order

The goal of the game is get the longest line of fractions listed in order from least to greatest using the following rules:

- All 20 fraction cards are placed faceup in any order. This is the selection pot.

- Players take turns selecting one card from the pot. This card is placed in the player's own line of cards so that the cards are in order from least to greatest.

- If a player disagrees with the placement of an opponent's card, he or she can challenge the placement. Together, the players should decide the correct placement. If the original placement was incorrect, the challenging player can take the card and put it in his or her line or return it to the selection pot.

- The game ends when all cards in the pot have been played. The winner is the player with the greater number of cards in his or her line.

Write About Comparing Fractions

Write an explanation of how to compare two fractions. Give examples to support your explanation.

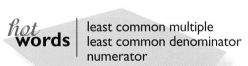

hot **words** | least common multiple
least common denominator
numerator

page 150

4/6

1/4   2/3

2/8

Now that you are familiar with so many aspects of fractions, it is time to add and subtract. You will start by using number lines. When you put fraction strips on the number line, the sums and differences make sense. Soon you will be adding and subtracting fractions without models. As you play games and solve unusual problems, you will become familiar with fractions greater than one.

Adding Parts and Taking Them Away

WHAT'S THE MATH?

Investigations in this section focus on:

NUMBERS and OPERATIONS

- Using a number line to model addition and subtraction of fractions
- Adding and subtracting fractions without using a model
- Converting between mixed numbers and improper fractions
- Adding and subtracting mixed numbers and improper fractions

MathScape Online
mathscape1.com/self_check_quiz

10 Sums and Differences on the Line

ADDING AND SUBTRACTING FRACTIONS ON THE NUMBER LINE

You have already learned to add and subtract whole numbers using a number line. In this lesson, you will use number lines to add and subtract fractions.

Use a Number Line to Add Fractions

How can you add fractions on a number line?

Each number line you use today is divided into twelfths.

1 Use number lines to find each sum. Then write a fraction that represents the sum.

a. $\frac{1}{3} + \frac{1}{4}$ b. $\frac{2}{3} + \frac{1}{4}$ c. $\frac{1}{4} + \frac{5}{12}$

d. $\frac{7}{12} + \frac{1}{2}$ e. $\frac{5}{6} + \frac{3}{4}$ f. $\frac{1}{6} + \frac{3}{4}$

2 Choose one of your addition problems. How did you determine the denominator for your answer?

Adding Fractions Using Number Lines

To show $\frac{1}{2}$, make a paper strip that fills $\frac{1}{2}$ of the space between 0 and 1.

To show $\frac{1}{3}$, make a paper strip that fills $\frac{1}{3}$ of the space between 0 and 1.

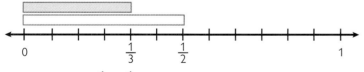

To find the sum of $\frac{1}{2} + \frac{1}{3}$, place the strips end to end with zero as the starting point. Find their total length. What fraction represents the sum?

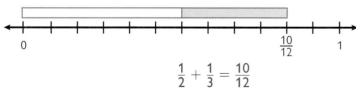

$$\frac{1}{2} + \frac{1}{3} = \frac{10}{12}$$

Use a Number Line to Subtract Fractions

Now that you can add fractions on the number line, try subtracting them. To subtract, find how much greater one fraction strip is than the other.

1 Use number lines to find each difference. Write a fraction that represents each difference.

a. $\frac{5}{6} - \frac{1}{4}$ **b.** $\frac{3}{4} - \frac{1}{3}$

c. $\frac{1}{2} - \frac{1}{6}$ **d.** $\frac{11}{12} - \frac{1}{6}$

e. $\frac{2}{3} - \frac{7}{12}$ **f.** $\frac{1}{2} - \frac{1}{3}$

2 Choose one of your subtraction problems. Explain how you found the difference. How did you determine the denominator for your answer?

3 Find each sum or difference.

a. $\frac{1}{3} - \frac{1}{6}$ **b.** $\frac{3}{4} - \frac{1}{6}$

c. $\frac{1}{2} + \frac{3}{4}$ **d.** $\frac{5}{6} + \frac{1}{3}$

e. $\frac{11}{12} - \frac{2}{3}$ **f.** $\frac{5}{6} - \frac{1}{4}$

g. $\frac{2}{3} + \frac{5}{6}$ **h.** $\frac{1}{3} + \frac{1}{4}$

i. $1 - \frac{5}{6}$ **j.** $\frac{7}{12} - \frac{1}{2}$

Write a Definition

In your copy of My Math Dictionary, write a definition of *common denominator*, and give an example.

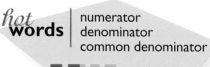
hot **words** | numerator
denominator
common denominator

Homework
page 151

11 Numbers Only

ADDING AND
SUBTRACTING
FRACTIONS
WITHOUT MODELS

Using area models or number lines is not always the simplest way to add and subtract fractions. In this lesson, you will devise a way to find sums and differences of fractions without these tools.

Add and Subtract Fractions Without Models

How can you add and subtract fractions without using models?

Use what you have learned about models for adding and subtracting fractions to devise a method to add and subtract fractions without models.

1 Copy and complete the addition table. Find each sum and describe how you found the sum.

	Problem	Answer	Description of Method
a.	$\frac{2}{5} + \frac{1}{5}$		
b.	$\frac{1}{2} + \frac{1}{8}$		
c.	$\frac{3}{8} + \frac{1}{4}$		
d.	$\frac{7}{10} + \frac{3}{5}$		

2 What steps would you use to find the sum of any two fractions?

3 Copy and complete the subtraction table. Find each difference and describe how you found the difference.

	Problem	Answer	Description of Method
a.	$\frac{6}{7} - \frac{2}{7}$		
b.	$\frac{2}{3} - \frac{4}{9}$		
c.	$\frac{7}{12} - \frac{1}{4}$		
d.	$\frac{3}{5} - \frac{2}{15}$		

4 What steps would you use to find the difference between any two fractions?

Play "Race for the Wholes"

The game "Race for the Wholes" can be played with 2, 3, or 4 players. Play the game with some of your classmates.

Can you make a runner land on a whole number?

Race for the Wholes

To play this game, you will need a set of fraction cards, a number line, and the same number of game markers (racers) as players. The number line should be marked with the numbers from 0 to 3 with 24 spaces between whole numbers.

- Begin the game with all racers on 0.

- Each player is dealt 3 cards and the rest remain facedown in a pile.

- Players take turns. On each turn, the player moves any racer by using one of his or her fraction cards. The player can move the racer forward or backward the distance shown on the card.

- After playing a card, the player puts it in a discard pile and draws a new card to replace it.

- A player earns one point each time he or she can make a racer land on a whole number.

- The first player with 5 points is the winner.

Solve the Magic Square

In a magic square, the sum of the numbers in each column, row, and diagonal is the same. Arrange the numbers into the nine boxes so that they form a magic square. The sum of the fractions in each column, row, and diagonal will be 1.

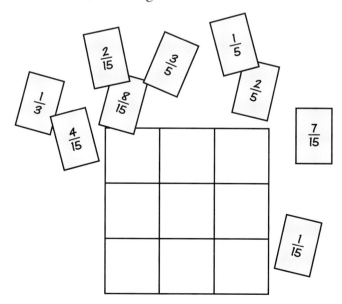

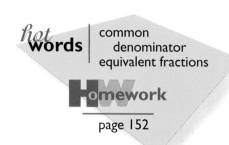

hot **words** | common denominator
equivalent fractions

H🔲**mework**

page 152

12 Not Proper but Still Okay

ADDING MIXED
NUMBERS AND
IMPROPER
FRACTIONS

What do you do with fractions that are greater than one?
How you choose to write these numbers usually depends on what you are doing with them. But, do not be misled by the names. An improper fraction is not bad, and a mixed number is not confused!

Write Mixed Numbers and Improper Fractions

How are mixed numbers and improper fractions related?

Numbers that are not whole numbers and are greater than one can be written as mixed numbers or as improper fractions.

1 Draw an area model or number line that shows each improper fraction. Then, write the corresponding mixed number.

a. $\frac{4}{3}$ b. $\frac{7}{4}$ c. $\frac{12}{5}$ d. $\frac{7}{2}$

2 Draw an area model or number line that shows each mixed number. Then, write the corresponding improper fraction.

a. $2\frac{1}{3}$ b. $4\frac{1}{4}$ c. $3\frac{3}{5}$ d. $1\frac{5}{6}$

3 Write each number in a different form without using an area model or a number line.

a. $3\frac{4}{5}$ b. $\frac{15}{5}$ c. $\frac{10}{4}$ d. $6\frac{2}{3}$

4 Explain how you can convert between improper fractions and mixed numbers without using area models or number lines.

Improper Fractions and Mixed Numbers

A fraction with a numerator less than the denominator is a **proper fraction.**

A fraction with a numerator greater than or equal to the denominator is an **improper fraction.**

A **mixed number** is a mix of a whole number and a proper fraction.

Both the area model and the number line show the same number. It can be called $\frac{5}{3}$ or $1\frac{2}{3}$.

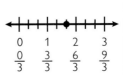

Add Fractions Greater than One

You can use your knowledge of adding fractions to help you add mixed numbers and improper fractions.

How can you add mixed numbers and improper fractions?

1 With a partner or group, find at least two ways to add the following numbers. If the problem is written with improper fractions, write the sum as an improper fraction. If the problem is written with mixed numbers, write the sum as a mixed number. If the problem has both types of numbers, write the sum as an improper fraction and as a mixed number.

a. $\frac{4}{3} + \frac{5}{6}$ 　　　　　　　　　　**b.** $\frac{3}{2} + \frac{6}{5}$

c. $2\frac{1}{2} + 3\frac{3}{4}$ 　　　　　　　　　**d.** $4\frac{1}{3} + 1\frac{2}{5}$

e. $2\frac{1}{4} + \frac{5}{2}$ 　　　　　　　　　**f.** $\frac{4}{3} + 3\frac{1}{3}$

2 With a partner, discuss two different methods of adding mixed numbers and improper fractions. Is one method better or more efficient than the other? Explain your reasoning. Be ready to share your methods with the class.

Mixed Number Rule

Mixed numbers should not include an improper fraction. So, you sometimes need to rename the fraction.

For example, if you add $2\frac{3}{5}$ and $4\frac{4}{5}$, you might find that the sum is $6\frac{7}{5}$. But, you cannot have an improper fraction in a mixed number. So, your next step is to rename the improper fraction within the mixed number.

Since $\frac{7}{5} = 1\frac{2}{5}$, $6\frac{7}{5}$ is the same as $6 + 1 + \frac{2}{5}$.

So, you should write $6\frac{7}{5}$ as $7\frac{2}{5}$.

Write About Fractions

In your copy of My Math Dictionary, describe and give examples of *improper fraction* and *mixed number*.

hot words ┃ improper fractions
mixed numbers

Homework

page 153

13 Sorting Out Subtraction

Now that you have added mixed numbers and improper fractions, it is time to consider subtraction. You can use what you know about adding mixed numbers and improper fractions to help you find ways to subtract these types of numbers.

Subtract Fractions Greater than One

How can you subtract mixed numbers and improper fractions?

The information below about regrouping in subtraction will help you subtract mixed numbers.

1 Find each difference.

a. $\frac{10}{9} - \frac{2}{9}$ b. $\frac{25}{12} - \frac{7}{6}$

c. $3 - 1\frac{2}{7}$ d. $5\frac{3}{4} - 2\frac{1}{7}$

e. $4\frac{11}{12} - 3\frac{3}{4}$ f. $5 - 2\frac{4}{5}$

g. $6\frac{2}{3} - \frac{9}{2}$ h. $\frac{21}{5} - 2\frac{3}{10}$

2 When you have found the answers to all eight problems, be ready to explain the steps you used to get each answer.

Regrouping in Subtraction

When you subtract mixed numbers, you sometimes need to regroup. That is, you need to take one whole from the whole number part and add it to the fraction part.

To find $5\frac{1}{12} - \frac{17}{12}$, use the following steps:

- First, regroup.

  $5\frac{1}{12} = 4 + 1 + \frac{1}{12}$ *Take one whole from the 5.*

  $= 4 + \frac{12}{12} + \frac{1}{12}$ *Rename the whole as $\frac{12}{12}$.*

  $= 4 + \frac{13}{12}$ *Add $\frac{12}{12}$ and $\frac{1}{12}$.*

- Then, subtract.

  $4\frac{13}{12} - 1\frac{7}{12} = 3\frac{6}{12}$ or $3\frac{1}{2}$

Get That Robot Out of Here!

Against the back wall of your room there is a large robot with an annoying bug in its programming. The robot has to travel 10 meters forward to get through the door, and you need to get it out as soon as possible. You can tell the robot to move toward the door using any of the given distances listed below. However, because of the bug, on every other move the robot will move in the wrong direction. So, if you tell the robot to go $\frac{1}{2}$ meter and then tell it to go $\frac{1}{3}$ meter, it will go forward $\frac{1}{2}$ meter and then back $\frac{1}{3}$ meter. The robot will only be $\frac{1}{6}$ meter closer to the door.

To get the robot out, you can choose any of the numbers below, but you can use each number only once. Create a table like the one below to record your progress. Notice the *distance from wall* plus *distance to door* must always equal 10 meters.

What is the least number of moves that you can use to get the robot out of the room?

What is the least number of computations that will get the robot out of the room?

Move Number	Add or Subtract	Distance from Wall	Distance to Door
1	$+2\frac{1}{3}$	$2\frac{1}{3}$	$7\frac{2}{3}$
2	$-\frac{1}{4}$	$2\frac{1}{12}$	$7\frac{11}{12}$

Possible Moves

Each move can be used only once. All distances are in meters.

$\frac{1}{2}$	$\frac{1}{4}$	$\frac{4}{3}$	$\frac{13}{6}$	$\frac{5}{6}$
$\frac{3}{2}$	$\frac{9}{4}$	$2\frac{1}{3}$	$\frac{1}{12}$	$\frac{11}{12}$
$\frac{1}{6}$	$\frac{7}{2}$	$\frac{3}{4}$	$\frac{5}{12}$	$1\frac{1}{6}$
$\frac{5}{3}$	$\frac{7}{3}$	$\frac{4}{4}$	$\frac{1}{3}$	$2\frac{1}{4}$

hot **words** | improper fraction
mixed number

page 154

14 Calc and the Numbers

DESCRIBING AND USING RULES FOR FRACTION ADDITION AND SUBTRACTION

An extraterrestrial visitor named Calc has landed in your classroom. He likes math, but so far he can only work with whole numbers. You are going to teach him about fractions. Fortunately, he is excellent at following directions.

Write Directions for Adding Fractions

Can you write directions to explain how to add fractions?

Calc is from the planet Integer. He can only work with whole numbers. He follows directions perfectly, and he understands terms such as *numerator* and *denominator*.

1 Write step-by-step instructions that tell Calc how to add two fractions. He needs to learn how to add:

a. fractions with like denominators,

b. fractions with unlike denominators,

c. improper fractions, and

d. mixed numbers.

2 Test your instructions on the problems below. Would Calc be able to find all of the answers using your instructions?

a. $\frac{3}{7} + \frac{2}{7}$ **b.** $\frac{1}{3} + \frac{1}{4}$ **c.** $\frac{9}{5} + \frac{11}{5}$

d. $\frac{5}{3} + \frac{14}{12}$ **e.** $7\frac{1}{2} + 4\frac{1}{3}$ **f.** $2\frac{3}{4} + 1\frac{5}{8}$

Use Fractions

Calc returned home and told his fellow Integerlings the news about fractions. Now, some Integerlings need to use fractions. See if you can answer each of their problems.

Can you add and subtract fractions to solve problems?

1 Maribeth is planning to carpet two rooms with matching carpet.

   a. Both rooms are 12 feet wide. One room is $9\frac{3}{4}$ feet long. The other is $8\frac{2}{3}$ feet long. She finds some carpet that is 12 feet wide. How long of a piece of carpet does she need?

   b. Maribeth decides she should have some extra carpet in case she makes a mistake installing it. She buys a 20-foot length. She does a perfect job installing the carpet. How much carpet is left over?

2 Ronita plans to replace the floor molding around the perimeter of a room.

   a. The room measures $12\frac{1}{4}$ feet by 11 feet. It has one doorway, which is $3\frac{1}{4}$ feet wide. If Ronita does not put molding along the doorway, what is the total length of molding she needs?

   b. Ronita's brother gives her $30\frac{1}{2}$ feet of molding. How much more molding does Ronita need?

3 John is a tailor and is making some suits.

   a. He has $12\frac{1}{4}$ yards of fabric. He needs $4\frac{2}{3}$ yards for one customer and $5\frac{3}{4}$ yards for another. How much fabric does he need in all? Does he have enough fabric?

   b. John's customers both decided to order vests, so he needs $\frac{3}{4}$ yard more of the fabric for each customer's suit. How much fabric does he need now? How much will he have leftover when he is done?

hot **words** | equivalent fractions
common denominator

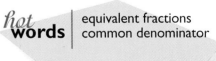

page 155

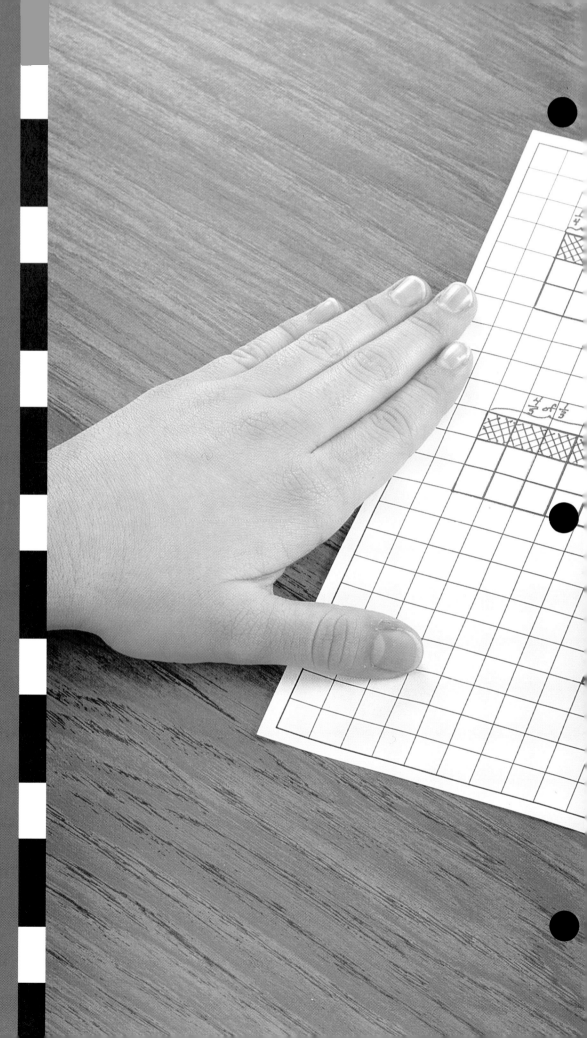

As you will see in this phase, fraction multiplication and division are easy to compute. But, why do they work the way they do and how do they relate to each other? Think of this phase as a study in group behavior – of fractions! By the end of the phase, you will be playing fraction multiplication and division games. With your new understanding of fraction groups, you will be able to estimate and make predictions about problems that used to seem mysterious.

Fractions in Groups

WHAT'S THE MATH?

Investigations in this section focus on:

NUMBERS and OPERATIONS

- Finding and applying methods to multiply a whole number by a fraction

- Using an area model to find a fraction of a fraction

- Finding and applying a method for multiplying fractions

- Estimating products

- Using models to divide by fractions

- Finding and applying a method for dividing fractions

MathScape Online
mathscape1.com/self_check_quiz

15 Picturing Fraction Multiplication

FINDING FRACTIONS
OF WHOLES

How can you use drawings to multiply a fraction by a whole? In this lesson, you will learn several ways to multiply using drawings. Which drawings work best for you may depend on how you think about the problems.

Find Fractions of Wholes

How can you use a drawing to find a fraction of a whole number?

When a math class was asked to show $\frac{1}{3}$ of 3, the students made a variety of drawings. How does Maya's drawing show that $\frac{1}{3}$ of 3 is 1? How does David's drawing show that $\frac{1}{3}$ of 3 is 1?

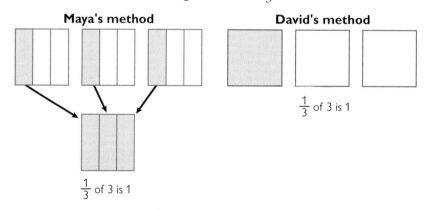

Maya's method

David's method

$\frac{1}{3}$ of 3 is 1

$\frac{1}{3}$ of 3 is 1

1 Solve each problem in the handout Reproducible R18.

2 Use Maya's method, David's method, or a method of your choice to solve each problem.

　　a. $\frac{5}{6}$ of 12　　　　　　b. $\frac{2}{3}$ of 6

　　c. $\frac{3}{5}$ of 25　　　　　　d. $\frac{1}{3}$ of 18

　　e. $\frac{5}{8}$ of 40　　　　　　f. $\frac{5}{9}$ of 3

3 How did you solve each problem in part **2**? Did you use the same method for every problem? Are some problems easier to solve using Maya's method than David's method?

Use Number Lines to Find Fractions of Wholes

How can you use a number line to find a fraction of a whole number?

To find $\frac{2}{3}$ of 6 on a number line, first draw a number line 6 units long. Then, divide the line into three equal sections. Each section is one third of the line. To find $\frac{2}{3}$ of 6, you will need two sections. So, $\frac{2}{3}$ of 6 is 4.

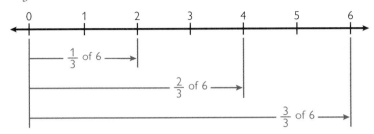

1 Use a number line to solve each problem.

a. $\frac{4}{5}$ of 20

b. $\frac{5}{6}$ of 36

c. $\frac{2}{9}$ of 18

d. $\frac{1}{6}$ of 24

e. $\frac{3}{4}$ of 16

f. $\frac{2}{5}$ of 10

2 Do you prefer to use drawings or number lines to find fractions of whole numbers? Explain.

Compute a Fraction of a Whole Number

After Kary and Jesse multiplied fractions using drawings and number lines, each of them devised a plan to multiply without using models.

Kary's Method	Jesse's Method
First, multiply the whole number by the numerator of the fraction. Then, divide the result by the denominator of the fraction.	First, divide the whole number by the denominator of the fraction. Then, multiply the result by the numerator of the fraction.

1 Use both Kary's method and Jesse's method to solve each problem from part 1 above using the number line.

2 Decide whether Kary, Jesse, or both are correct. If both are correct, which do you prefer?

3 Find $\frac{1}{6}$ of 15. Explain how you found your answer.

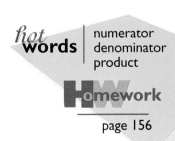

hot **words**
numerator
denominator
product

Homework
page 156

16 Fractions of Fractions

MULTIPLYING
FRACTIONS

In this lesson, you will find fractions of fractions. When you find a fraction of a fraction, you are actually multiplying the numbers. Making rectangular models will help you find a method to multiply fractions.

Finding Fractions of Fractions

How can you find a fraction of a fraction?

You can use a rectangular model to multiply fractions.

1 For each problem, the rectangle represents one whole. Use the rectangle to find each solution. Write a multiplication statement for each problem.

a. $\frac{1}{4}$ of $\frac{1}{3}$

b. $\frac{1}{2}$ of $\frac{1}{4}$

c. $\frac{4}{5}$ of $\frac{1}{3}$

d. $\frac{2}{7}$ of $\frac{2}{3}$

2 Study the numerators and denominators in the multiplication statements you wrote. Write a method for multiplying fractions.

Test a Hypothesis

Study the results of a variety of multiplication problems to determine how the first factor affects the product.

When does a multiplication problem result in a product that is less than a factor?

1 Make a table like the one below. Then, complete the table using the given problems. The factor type can be *whole number*, *improper fraction*, or *proper fraction*.

Problem	Answer	First Factor Type	Is Answer Greater or Less Than the Second Factor?
3×5	15	whole number	greater than
$\frac{3}{4} \times 12$	9	proper fraction	less than

a. $12 \times \frac{3}{4}$　　　**b.** $\frac{2}{3} \times 9$　　　**c.** $9 \times \frac{2}{3}$

d. $\frac{5}{4} \times 20$　　　**e.** $10 \times \frac{3}{2}$　　　**f.** $\frac{1}{4} \times \frac{1}{4}$

g. $\frac{4}{9} \times \frac{1}{3}$　　　**h.** $\frac{1}{3} \times \frac{4}{3}$　　　**i.** $\frac{3}{2} \times \frac{4}{3}$

j. $\frac{5}{4} \times \frac{12}{10}$　　**k.** $0 \times \frac{2}{3}$　　　**l.** $6 \times \frac{5}{4}$

2 Highlight each row that has a *less than* in the last column. Use a different color to highlight each row that has a *greater than* in the last column.

3 Analyze your results. What type of factor(s) makes the product greater than the other number you were multiplying? What type of factor(s) makes the product less than the other number you were multiplying? Why do you think this is true?

Multiplying Fractions

To multiply two fractions use the following procedure.

$$\frac{2}{3} \times \frac{3}{5}$$

$\frac{2 \times 3}{3 \times 5} \rightarrow \frac{}{15}$　　Multiply the denominators. This is like dividing each fifth into 3 pieces. Each piece is a fifteenth.

$\frac{2 \times 3}{3 \times 5} \rightarrow \frac{6}{15}$　　Multiply the numerators. The numerators tell how many pieces you want. You have 3 fifths, and you want 2 of the 3 pieces in each fifth. You want a total of 6 pieces.

hot **words** | numerator
denominator
improper fraction

page 157

17 Estimation and Mixed Numbers

ESTIMATING
RESULTS AND
MULTIPLYING
MIXED NUMBERS

Estimating is an important skill. In this lesson, you will estimate fraction products. As you get a better sense of how numbers work in fraction multiplication, you will be better at estimating fractions.

Estimate Fraction Products

Can you estimate fraction products?

For each multiplication problem, choose the estimate that you think will be the closest to the actual answer. Then, play "Best Estimate" with your classmates.

1 $\frac{10}{5} \times \frac{3}{4}$ **a.** $\frac{1}{2}$ **b.** 3   **c.** $1\frac{1}{2}$

2 $\frac{5}{4} \times 13$ **a.** 13   **b.** 15   **c.** 10

3 $\frac{1}{5} \times 26$ **a.** 100   **b.** $5\frac{4}{5}$ **c.** 5

4 $11 \times \frac{1}{5}$ **a.** 2   **b.** $2\frac{1}{2}$ **c.** 55

5 $\frac{3}{2} \times 48$ **a.** 16   **b.** 75   **c.** 148

Best Estimate

Get Ready

- Each student makes up 3 multiplication problems and solves the problems.
- Each student makes up 3 possible answers for each problem. One of the answers is close to the actual answer. The other two answers are not.
- Each student writes each problem plus its 3 answers on an index card.

Play the Game

- Divide into groups. Each student takes a turn presenting a problem.
- The other group members have 10 seconds to decide which of the answers is the closest estimate and write the choice on a piece of paper.
- All students who choose the best estimate receive a point.

Multiply Mixed Numbers

You know how to multiply fractions and whole numbers and how to multiply fractions and fractions. How can you multiply with mixed numbers?

How can you multiply mixed numbers?

1 Study the equations involving mixed numbers. Try to find a method or methods for multiplying mixed numbers. Be sure your methods work for all the examples.

$$1\frac{1}{3} \times 2 = \frac{8}{3} \text{ or } 2\frac{2}{3} \qquad \frac{4}{5} \times 3\frac{2}{5} = \frac{68}{25} \text{ or } 2\frac{18}{25}$$

$$2\frac{1}{2} \times 3 = \frac{15}{2} \text{ or } 7\frac{1}{2} \qquad \frac{1}{2} \times 1\frac{1}{2} = \frac{3}{4}$$

$$4 \times 1\frac{1}{8} = \frac{36}{8} \text{ or } 4\frac{4}{8} \text{ or } 4\frac{1}{2} \qquad 1\frac{1}{2} \times 1\frac{1}{2} = \frac{9}{4} \text{ or } 2\frac{1}{4}$$

$$2\frac{1}{2} \times \frac{1}{2} = \frac{5}{4} \text{ or } 1\frac{1}{4} \qquad 1\frac{1}{3} \times 1\frac{1}{4} = \frac{20}{12} \text{ or } 1\frac{8}{12} \text{ or } 1\frac{2}{3}$$

2 Write a method or methods for multiplying mixed numbers. Be prepared to explain your methods to your teacher and classmates.

Write about Multiplying Mixed Numbers

What happens when you multiply a number by a mixed number? Is the product greater or less than the other number? Before writing your conclusion, make sure you try multiplying a mixed number by:

- proper fractions,
- improper fractions,
- whole numbers, and
- mixed numbers.

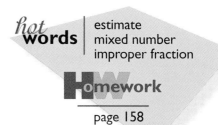

hot **words** estimate
mixed number
improper fraction

Homework
page 158

18 Fraction Groups within Fractions

DIVIDING WITH
FRACTIONS

You know how to multiply a number by a fraction.
What happens when you divide a number by a fraction? What you have learned about multiplying fractions will be helpful when you divide fractions.

Find Fraction Groups in Whole Numbers

How can you divide a whole number by a fraction?

Here are two ways to model the problem $6 \div \frac{1}{2}$.

There are 12 halves in 6.

$$6 \div \frac{1}{2} = 12$$

1 Make either a drawing or a number line to find each quotient.

a. $6 \div \frac{1}{3}$ **b.** $6 \div \frac{1}{4}$ **c.** $6 \div \frac{1}{5}$

2 Look at the problems and answers for part **1.**

   a. Describe a method that you could use to find the answers.

   b. Use your method to find $8 \div \frac{1}{3}$ and $6 \div \frac{1}{6}$.

3 To find each quotient, refer to part **1.**

   a. Use the drawing or number line for part **1a** to find $6 \div \frac{2}{3}$.

   b. Use the drawing or number line for part **1b** to find $6 \div \frac{3}{4}$.

   c. Use the drawing or number line for part **1c** to find $6 \div \frac{3}{5}$.

4 Look at the problems and answers for part **3.**

   a. Describe a method that you could use to find the answers.

   b. Use your method to find $8 \div \frac{2}{3}$ and $6 \div \frac{5}{6}$.

Divide Fractions and Mixed Numbers

Use what you have learned about fraction division to find each quotient.

1 $9 \div \frac{2}{3}$ **2** $\frac{4}{5} \div 9$ **3** $\frac{5}{6} \div \frac{8}{9}$

4 $\frac{3}{4} \div \frac{1}{2}$ **5** $\frac{7}{10} \div \frac{4}{5}$ **6** $\frac{1}{2} \div \frac{1}{4}$

7 $\frac{9}{5} \div \frac{2}{3}$ **8** $\frac{5}{9} \div \frac{1}{4}$ **9** $1\frac{1}{2} \div \frac{1}{8}$

10 $9 \div \frac{4}{3}$ **11** $12 \div \frac{5}{4}$ **12** $3\frac{3}{4} \div \frac{3}{8}$

How can you divide fractions?

Dividing Fractions

When you divide a number by a fraction, the denominator of the fraction breaks the number into more parts. So, you multiply by the denominator. Then, the numerator of the fraction tells you to regroup these parts. So, you divide by the numerator.

You can use this method to find $6 \div \frac{3}{4}$:

$6 \times 4 = 24$ This tells you there are a total of 24 parts.

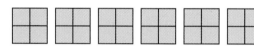

$24 \div 3 = 8$ There are 8 groups of 3 parts.

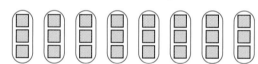

A shortcut for dividing with fractions is to rewrite the division problem as a number multiplied by the **reciprocal** of the divisor.

$$6 \div \frac{3}{4} = 6 \times \frac{4}{3} = \frac{6}{1} \times \frac{4}{3} = \frac{24}{3} \text{ or } 8$$

This shortcut works with any division problem.

$14 \div 7 = \frac{14}{1} \div \frac{7}{1} = \frac{14}{1} \times \frac{1}{7} = \frac{14}{7} \text{ or } 2$

$\frac{2}{3} \div \frac{3}{4} = \frac{2}{3} \times \frac{4}{3} = \frac{8}{9}$

$\frac{5}{6} \div 2 = \frac{5}{6} \div \frac{2}{1} = \frac{5}{6} \times \frac{1}{2} = \frac{5}{12}$

$1\frac{1}{2} \div \frac{6}{5} = \frac{3}{2} \div \frac{6}{5} = \frac{3}{2} \times \frac{5}{6} = \frac{15}{12} \text{ or } 1\frac{1}{4}$

hot words | reciprocal

Homework
page 159

19 Understanding Fraction Division

What trends can you discover in division with fractions? In this lesson, you will use a familiar investigation to learn more about division by fractions.

Explore the Effects of Division

What happens when a number is divided by a number between 0 and 1?

Sue said, "We have already learned that when you multiply a number by a number that is greater than one, the product is greater than the other number. When you multiply by a proper fraction, the product is less than the other number. I think the opposite is true for division."

Is Sue correct? The following investigation will help you decide.

1 Make a table like the one below. Then, complete the table using the given problems. The divisor type can be *whole number, improper fraction, proper fraction,* or *mixed number.*

Problem	Answer	Divisor Type	Is Answer Greater or Less Than the Dividend?
$10 \div 2$	5	whole number	less than
$10 \div \frac{1}{2}$	20	proper fraction	greater than

a. $\frac{3}{4} \div 4$ b. $\frac{5}{12} \div 6$ c. $\frac{1}{2} \div \frac{1}{4}$

d. $\frac{1}{8} \div \frac{1}{3}$ e. $\frac{4}{3} \div \frac{1}{3}$ f. $\frac{2}{9} \div \frac{5}{4}$

g. $\frac{5}{8} \div \frac{3}{2}$ h. $1\frac{1}{2} \div \frac{1}{8}$ i. $\frac{4}{5} \div 3\frac{1}{4}$

j. $2\frac{1}{4} \div 1\frac{1}{2}$ k. $5 \div 1\frac{1}{3}$ l. $\frac{5}{9} \div \frac{5}{6}$

2 Analyze your results. What type(s) of divisors make the quotient greater than the dividend? What type(s) of divisors make the quotient less than the dividend? Why do you think this is true?

Estimate the Quotients

For each division problem, choose the estimate that you think will be closer to the real answer. Then play "Better Estimate" with your classmates.

Can you estimate fraction quotients?

1 $4 \div \dfrac{1}{8}$ **a.** 32   **b.** $\dfrac{1}{2}$

2 $3\dfrac{1}{2} \div \dfrac{1}{3}$ **a.** 21   **b.** 10

3 $4 \div \dfrac{5}{4}$ **a.** $4\dfrac{1}{4}$ **b.** 3

4 $\dfrac{5}{9} \div \dfrac{1}{4}$ **a.** 2   **b.** $\dfrac{1}{2}$

5 $25 \div \dfrac{5}{6}$ **a.** 23   **b.** 30

6 $\dfrac{1}{2} \div \dfrac{25}{12}$ **a.** $\dfrac{1}{4}$ **b.** 20

Better Estimate

Get Ready
- Each student makes up three division problems and solves the problems. The problems should include whole numbers, proper fractions, improper fractions, and mixed numbers.

- Each student makes up two possible answers for each problem. One of the answers is close to the actual answer. The other answer is an inaccurate guess or poor estimate.

- Each student writes each problem plus the two answers on an index card or quarter-sheet of paper.

Play the Game
- Divide into groups. Each student takes a turn presenting a problem.

- The other group members have 15 seconds to decide which of the answers is the closer estimate and write the choice on a piece of paper.

- When time is up, everyone holds up his or her answer choice. All students who choose the better estimate receive a point.

- The player with the most points after all problems have been presented wins.

hot **words** | mixed number
improper fraction
estimate

 omework

page 160

Multiplication vs. Division

APPLYING FRACTION
MULTIPLICATION
AND DIVISION

Now, you are familiar with how to multiply and divide fractions. In this lesson, you will apply what you have learned about fractions in a game that combines both operations.

Play "Get Small"

How can you multiply and divide fractions to get the least answer?

Play a round of "Get Small" on your own or with a partner. Make a recording sheet like the one below to show your work.

First Fraction	Second Fraction	× or ÷	Resulting Fraction
$\frac{5}{4}$	$\frac{3}{4}$	×	$\frac{15}{16}$
$\frac{15}{16}$	$\frac{3}{2}$	÷	$\frac{30}{48}$ or $\frac{5}{8}$

Get Small

In this game, the player rolls a number cube to create fractions. These fractions are multiplied or divided to create the least possible number. Use the following rules to play the game.

- The player rolls the number cube and places the result in either the numerator or the denominator of the first fraction.
- Then, the player rolls the number cube to determine the other part of the first fraction.
- The player rolls two more times to create a second fraction.
- The player decides whether to multiply or divide the two fractions. The result becomes the first fraction for the next round.
- As before, the player rolls twice to create the second fraction for the next round. Then, he or she decides to multiply or divide.
- The player continues the game for 10 rounds. The object of the game is to end with the least possible number.

Find the Relationship between Multiplication and Division

You and your partner will solve two similar sets of problems. One of you will work on multiplication and the other will work on division. When you have both finished, compare your work. What relationships do you see between the problems?

How are division and multiplication of fractions related?

Student A	Student B
1 $\frac{1}{2} \times \frac{1}{3}$	**1** $\frac{1}{6} \div \frac{1}{2}$
2 $\frac{5}{6} \times \frac{3}{4}$	**2** $\frac{5}{8} \div \frac{5}{6}$
3 $\frac{4}{3} \times \frac{1}{3}$	**3** $\frac{4}{9} \div \frac{4}{3}$
4 $1\frac{1}{5} \times \frac{2}{5}$	**4** $\frac{12}{25} \div 1\frac{1}{5}$
5 $5 \times \frac{1}{4}$	**5** $\frac{5}{4} \div \frac{1}{4}$
6 $\frac{7}{5} \times 3$	**6** $\frac{21}{5} \div 3$
7 $\frac{3}{8} \times \frac{1}{4}$	**7** $\frac{3}{32} \div \frac{1}{4}$
8 $\frac{5}{4} \times \frac{3}{2}$	**8** $\frac{15}{8} \div \frac{5}{4}$
9 $3\frac{1}{2} \times 4$	**9** $14 \div 4$
10 $\frac{4}{5} \times \frac{20}{3}$	**10** $\frac{16}{3} \div \frac{20}{3}$

Solve Real-Life Problems

Solve each problem using multiplication or division.

1 There are $3\frac{3}{4}$ pies left after Julia's Fourth of July party. Her family has 4 members. How much pie will each family member get if they divide the remaining pies evenly?

2 Joe earns $8.00 an hour mowing lawns. He has spent $5\frac{1}{4}$ hours mowing lawns this week. How much money did he earn?

3 Keisha reads 3 books each week. On average, how many books does she read in a day?

4 Claire has a CD that takes $\frac{3}{4}$ of an hour to play. How long will it take to play the CD 7 times?

hot **words** | numerator denominator

page 161

Shapes and Factors

Applying Skills

Determine how many rectangles can be made from each number of tiles. Give the dimensions for each rectangle.

1. 14 tiles **2.** 18 tiles

3. 23 tiles **4.** 25 tiles

The statements in items 5 and 6 are about the rectangle below. Complete each statement.

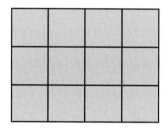

5. This rectangle's _____?_____ are 3 centimeters by 4 centimeters.

6. The two _____?_____ of 12 that this rectangle models are 3 and 4.

7. What two factors are common to every even number?

8. What factor is common to every number?

Suppose you are making solid rectangular prisms using 12, 15, 22, 27, 30, 40, and 48 cubes. Do not use 1 as a dimension.

9. Which numbers of cubes could form a rectangular solid that is 2 cubes high?

10. Which numbers of cubes could form a rectangular solid that is 3 cubes high?

Extending Concepts

Making a factor tree is one way to show the prime factorization of a number. Each branch of the tree shows the factors of the number above it. Two possible factor trees for the number 48 are shown below.

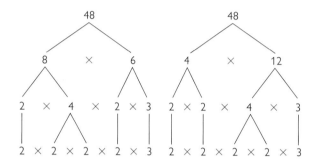

Make a factor tree for each number. Then write the number's prime factors.

11. 30 **12.** 81

13. 100 **14.** 66

15. 120 **16.** 250

Making Connections

17. Tamara wants to put a tiled patio in her backyard that is 8 feet by 12 feet. She plans to use 1-foot-square paving tiles. How many tiles does she need?

18. Masao built a raised planter box and needs to buy soil to put into it. The box is 4 feet wide and 8 feet long. He wants to fill it with soil to a depth of 2 feet. How many cubic feet of soil will he need?

The Great Factor Hunt

Applying Skills

Write each number as a product of its prime factors. Do not include the factor 1.

1. 12   **2.** 16   **3.** 21

4. 26   **5.** 30   **6.** 45

List all the factors other than 1 that each pair of numbers has in common.

7. 10 and 15   **8.** 8 and 19

9. 14 and 28   **10.** 6 and 24

11. 16 and 26   **12.** 40 and 50

Find the greatest common factor of each pair of numbers.

13. 4 and 6   **14.** 9 and 16

15. 12 and 24   **16.** 30 and 45

17. 20 and 50   **18.** 36 and 48

Extending Concepts

For items 19 to 24, use the number map below. Each time you move from one number to another, you multiply. You may move in any direction along the dotted line, but you may use each number only once. For example, the value of the highlighted path is $1 \times 5 \times 4 \times 11$ or 220.

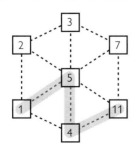

19. What is the longest path you can find? What is its value?

20. Can you find a path that is shorter than the one for item 19, but has the same value?

21. How can you make sure that a path's value will be an even number?

22. How can you make sure that a path's value will be an odd number?

23. Can you find a two-number path with a value that is prime? If possible, give an example.

24. Can you find a three-number path with a value that is prime? If possible, give an example.

Writing

25. Make up your own number map. In your map, include only prime numbers. Use the pattern below or make up your own. Write two questions to go with your map and give their answers.

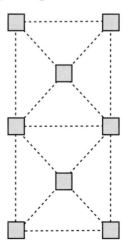

FROM WHOLES TO PARTS • HOMEWORK 2   **143**

Multiple Approaches

Applying Skills

Write the first five multiples of each number.

1. 3   **2.** 8

3. 13   **4.** 25

5. 40   **6.** 100

7. Determine whether 150 is a multiple of 6. Explain.

8. List three common multiples of 2 and 5.

9. List three common multiples of 3 and 9.

Find the least common multiple of each pair of numbers.

10. 6 and 8   **11.** 5 and 7

12. 9 and 12   **13.** 12 and 18

14. 9 and 21   **15.** 45 and 60

16. 11 and 17   **17.** 16 and 42

18. When will the least common multiple of two numbers be one of the numbers?

19. Suppose the only factor two numbers have in common is 1. How can you find their least common multiple?

Extending Concepts

To find the least common multiple of three numbers, first find the least common multiple of two of the numbers. Then find the least common multiple of that number and the third number.

Consider the numbers 10, 25, and 75. The least common multiple of 10 and 25 is 50. The least common multiple of 50 and 75 is 150. Therefore, the least common multiple of 10, 25, and 75 is 150.

Find the least common multiple of each set of numbers.

20. 16, 24, and 26   **21.** 15, 30, and 60

22. 4, 22, and 60   **23.** 14, 35, and 63

24. 9, 12, and 15   **25.** 15, 25, and 35

Making Connections

Our calendar year is based on the length of time Earth takes to travel around the Sun. The solar year is not exactly 365 days long. Therefore, every fourth year is a lcap year with 366 days. However, each solar year is not precisely $365\frac{1}{4}$ days. So, leap year is skipped every 400 years. Therefore, leap years are the years that are multiples of 4, but are not multiples of 400.

Determine which of the following years were leap years.

26. 1600   **27.** 1612

28. 1650   **29.** 1698

30. 1700   **31.** 1820

32. 1908   **33.** 2000

First Things First

Applying Skills

Solve for n in each equation.

1. $5 + 3 \times 7 = n$

2. $6 \div 3 + 4 = n$

3. $10 - 8 \div 2 + 3 = n$

4. $(6 - 3) \times 2 = n$

5. $3 \times 4 \div 6 + 2 = n$

6. $2 + 3 \times 4 = n$

7. $6 - 2 \times 3 = n$

8. $9 \div 3 + 4 = n$

9. $(3 \times 2)^2 - 10 = n$

10. $(3 + 6)^2 = n$

11. $5 \times (4 + 2)^2 = n$

12. $(5^2 + 3) \div 7 = n$

13. $13 + 6 - 4 + 12 = n$

14. $7 \times 9 - (4 + 3) = n$

15. $4 + 2(8 - 6) = n$

16. $3(4 + 7) - 5 \times 4 = n$

17. $19 + 45 \div 3 - 11 = n$

18. $9 + 12 - 6 \times 4 \div 2 = n$

19. $26 \div 13 + 9 \times 6 - 21 = n$

20. $27 - 8 \div 4 \times 3 = n$

Place each of the numbers 2, 5, and 6 into each equation so that the equation is true.

21. $\underline{\ ?\ } + \underline{\ ?\ } \times \underline{\ ?\ } = 32$

22. $\underline{\ ?\ } - \underline{\ ?\ } \div \underline{\ ?\ } = 2$

23. $\underline{\ ?\ } \times \underline{\ ?\ } \div \underline{\ ?\ } = 15$

24. $\underline{\ ?\ }^2 \div \underline{\ ?\ } - \underline{\ ?\ } = 13$

25. $\underline{\ ?\ }^2 \div \underline{\ ?\ }^2 + \underline{\ ?\ } = 14$

26. $(\underline{\ ?\ } + \underline{\ ?\ }) \times \underline{\ ?\ } = 22$

Extending Concepts

Imagine you are playing Hit the Target. For each set of building numbers, write an equation with the target number as the result.

27. building numbers: 1, 4, 5
target number: 2

28. building numbers: 2, 3, 6
target number: 1

29. building numbers: 3, 4, 6
target number: 2

30. building numbers: 3, 4, 6
target number: 6

Writing

31. Answer the letter to Dr. Math.

> Dear Dr. Math:
>
> My teacher gave our class a really easy problem to solve. Here is the problem.
>
> $$16 - 8 \div 4 + 3 = n$$
>
> Priscilla Pemdas got 17 for the answer—crazy, huh? Priscilla said my answer, which was 5 of course, was wrong. I know I added, subtracted, and divided correctly. But the teacher agreed with Priscilla! What is going on?
>
> D. Finite Disorder

Putting It All Together

Applying Skills

For each pair, tell whether expression A has the same answer as expression B.

1. **A.** $5 \times 6 + 11$
   B. $5 \times 11 + 6$

2. **A.** $48 \div 12 \div 2$
   B. $48 \div (12 \div 2)$

3. **A.** $(30 - 11) - 2$
   B. $30 - (11 - 2)$

4. **A.** $3 \times 5 \times 7$
   B. $7 \times 3 \times 5$

5. **A.** $50 \div 2 + 25$
   B. $50 \div 25 + 2$

6. **A.** 25×5
   B. $20 \div 5 + 5 \times 5$

7. **A.** $3 + 3 \times 10$
   B. $(3 + 3) \times 10$

Solve each problem.

8. Angie is thinking of two numbers. Their sum is 50. Each of the numbers has 10 as a factor. Both numbers are greater than 10. What are the numbers?

9. Julio has read two books so far this month. One book has twice as many pages as the other. All together, the two books have 300 pages. How many pages does each book have?

10. Mr. Robinson has two children. One child is 3 times as old as the other. The sum of their ages is 16. What are the ages of the children?

11. Atepa has two cats of different ages. The least common multiple of their ages is 24. The greatest common factor of their ages is 4. How old are his cats?

12. Marcus is thinking of three numbers. Their greatest common factor is 6. Their sum is 36. What are the numbers?

Extending Concepts

Debbie, Mariana, Eduardo, and Gary all love to read. For items 13–16, use the clues to determine how many pages each of them has read so far this month.

- Debbie has read 300 pages.
- The number of pages read by Eduardo and Debbie share a common factor, 50.
- Gary has read twice as many pages as Debbie.
- All together, the four students have read 1,425 pages.
- The least common multiple of the number of pages read by Mariana and Eduardo is 350.

13. Debbie
14. Mariana
15. Eduardo
16. Gary

Writing

17. Use the ages of some family members, friends, or fictitious people to create a "guess my number" problem. Your problem should include the terms *factor* and/or *multiple*. Write the problem and its solution.

Designer Fractions

Applying Skills

For items 1–6, use a grid.

1. Shade $\frac{1}{2}$ of a 4-by-4 rectangle. How many squares are shaded?

2. Shade $\frac{1}{4}$ of a 4-by-4 rectangle. How many squares are shaded?

3. Shade $\frac{1}{8}$ of a 4-by-4 rectangle. How many squares are shaded?

4. Shade $\frac{1}{3}$ of a 4-by-3 rectangle. How many squares are shaded?

5. Shade $\frac{1}{6}$ of a 4-by-3 rectangle. How many squares are shaded?

6. Shade $\frac{1}{12}$ of a 4-by-3 rectangle. How many squares are shaded?

Write a fraction that describes the part of the whole represented by each color.

7.

8.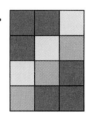

9.

10.

11.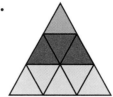

12.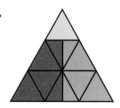

13.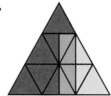

14.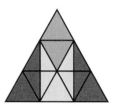

Extending Concepts

Answer each question.

15. If $\frac{1}{3}$ of a rectangle is shaded, what fraction of the rectangle is *not* shaded?

16. If $\frac{3}{7}$ of a rectangle is shaded, what fraction of the rectangle is *not* shaded?

Making Connections

17. The quilt pattern below is called Jacob's Ladder. Each block contains 9 same-sized squares. Each square is made of smaller squares or triangles. What fraction of the block is made of blue fabric? red fabric? white fabric?

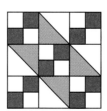

Area Models and Equivalent Fractions

Applying Skills

For items 1–4, use the 4-by-3 rectangle below.

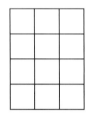

1. What fraction of the whole rectangle does one column represent?

2. What fraction of the whole rectangle do two rows represent?

3. Write two fractions with common denominators to represent one column and two rows.

4. Which fraction is greater?

For items 5–8, use the 2-by-7 rectangle below.

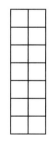

5. What fraction of the whole rectangle does one column represent?

6. What fraction of the whole rectangle do three rows represent?

7. Write two fractions with common denominators to represent one column and three rows.

8. Which fraction is greater?

Make a grid sketch for each pair of fractions. Label each sketch with the given fractions and their equivalent fractions. Compare the fractions.

9. $\frac{5}{8}$ and $\frac{3}{5}$

10. $\frac{5}{9}$ and $\frac{2}{3}$

11. $\frac{3}{4}$ and $\frac{6}{7}$

12. $\frac{3}{5}$ and $\frac{4}{7}$

For each pair of fractions, find equivalent fractions that have a common denominator. Compare the fractions.

13. $\frac{5}{12}$ and $\frac{1}{6}$

14. $\frac{2}{5}$ and $\frac{1}{4}$

15. $\frac{2}{7}$ and $\frac{3}{8}$

16. $\frac{1}{3}$ and $\frac{4}{7}$

Extending Concepts

Order each set of fractions from least to greatest. (*Hint:* To order fractions, rewrite all the fractions as equivalent fractions with a common denominator.)

17. $\frac{1}{2}, \frac{3}{5}, \frac{5}{6}$, and $\frac{2}{3}$

18. $\frac{1}{2}, \frac{4}{5}, \frac{2}{5}$, and $\frac{3}{4}$

Making Connections

19. Crystal's entire garden is a rectangle measuring 5 yards by 6 yards. She plans to plant a vegetable garden in $\frac{2}{3}$ of this area. How many square yards will she plant in vegetables? (Include a sketch with your solution.)

Fraction Lineup

Applying Skills

Make a number line like the one shown below. Place each fraction on the number line.

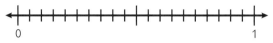

0   1

1. $\frac{1}{2}$

2. $\frac{3}{4}$

3. $\frac{4}{5}$

4. $\frac{3}{10}$

5. $\frac{14}{20}$

6. $\frac{3}{5}$

Write an equivalent fraction for each fraction.

7. $\frac{1}{2}$

8. $\frac{2}{10}$

9. $\frac{15}{20}$

10. $\frac{6}{10}$

11. $\frac{4}{5}$

12. $\frac{7}{10}$

Use a ruler to compare each pair of fractions. Write an expression using $<$, $>$ or $=$ to show the comparison.

IN.   1   2   3

13. $\frac{3}{8}$ and $\frac{1}{4}$

14. $\frac{8}{16}$ and $\frac{1}{2}$

15. $1\frac{1}{4}$ and $1\frac{3}{8}$

16. $2\frac{3}{4}$ and $2\frac{11}{16}$

17. $1\frac{1}{2}$ and $1\frac{9}{16}$

18. $\frac{3}{16}$ and $\frac{1}{8}$

Extending Concepts

19. Make a number line like the one shown below.

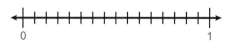

0   1

a. Label the number line with the fractions $\frac{1}{2}, \frac{1}{4}$, and $\frac{1}{8}$.

b. What would be the next fraction in the pattern? Label its position.

c. What would be the next two fractions in the pattern?

20. Make a number line like the one shown below. Then, locate each fraction.

8   9   10   11

a. $8\frac{1}{2}$

b. $10\frac{3}{4}$

c. $9\frac{1}{8}$

d. $8\frac{7}{8}$

Making Connections

Americans use fractions in daily life when measuring lengths and distances. Complete each sentence. Use 12 inches = 1 foot and 3 feet = 1 yard.

21. $1\frac{2}{3}$ yards is the same as __?__ feet.

22. 10 feet is the same as __?__ yards.

23. $\frac{1}{4}$ foot is the same as __?__ inches.

24. 18 inches is the same as __?__ feet.

25. $1\frac{1}{4}$ yards is the same as __?__ inches.

Focus on Denominators

Applying Skills

Find the least common denominator of each pair of fractions.

1. $\frac{2}{3}$ and $\frac{4}{15}$

2. $\frac{7}{8}$ and $\frac{3}{10}$

3. $\frac{7}{10}$ and $\frac{3}{5}$

4. $\frac{5}{7}$ and $\frac{2}{3}$

5. $\frac{7}{24}$ and $\frac{5}{12}$

6. $\frac{7}{8}$ and $\frac{5}{6}$

7. $\frac{3}{10}$ and $\frac{4}{15}$

8. $\frac{7}{12}$ and $\frac{5}{8}$

Compare each pair of fractions using >, < or =.

9. $\frac{4}{7}$ and $\frac{5}{8}$

10. $\frac{1}{3}$ and $\frac{3}{15}$

11. $\frac{4}{5}$ and $\frac{6}{7}$

12. $\frac{3}{8}$ and $\frac{5}{12}$

13. $\frac{3}{4}$ and $\frac{5}{8}$

14. $\frac{16}{20}$ and $\frac{40}{50}$

15. $\frac{5}{6}$ and $\frac{7}{9}$

16. $\frac{3}{16}$ and $\frac{1}{6}$

Order each set of fractions from least to greatest.

17. $\frac{2}{3}, \frac{5}{12},$ and $\frac{5}{6}$

18. $\frac{5}{16}, \frac{1}{4},$ and $\frac{9}{32}$

19. $\frac{32}{16}, \frac{1}{2},$ and $\frac{15}{16}$

20. $1\frac{1}{8}, \frac{5}{4},$ and $\frac{17}{16}$

Extending Concepts

Use each fraction once to correctly complete each statement.

$$\frac{5}{16} \quad \frac{41}{48} \quad \frac{37}{48} \quad \frac{17}{24}$$

21. $\frac{5}{6} < \underline{\quad ? \quad} < \frac{7}{8}$

22. $\frac{2}{3} < \underline{\quad ? \quad} < \frac{3}{4}$

23. $\frac{1}{4} < \underline{\quad ? \quad} < \frac{3}{4}$

24. $\frac{9}{12} < \underline{\quad ? \quad} < \frac{19}{24}$

Place each of the nine numbers as a numerator or denominator so that all of the equations are true. Use each of the nine numbers only once.

$$1 \quad 2 \quad 3 \quad 4 \quad 5 \quad 6 \quad 7 \quad 8 \quad 9$$

25. $\frac{?}{?} = \frac{?}{10}$

26. $\frac{2}{?} = \frac{?}{14}$

27. $\frac{2}{?} = \frac{?}{12}$

28. $\frac{8}{12} = \frac{?}{?}$

Writing

29. Answer the letter to Dr. Math.

> Dear Dr. Math:
> We are doing a unit about fractions, but our first phase was not even about fractions! It was just about whole-number stuff like factors and multiples. What do greatest common factors and least common multiples have to do with fractions anyway?
> Unclear N. Concept

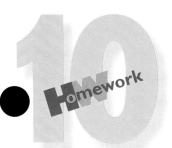

Sums and Differences on the Line

Applying Skills

Find each sum or difference. Use a number line if you wish.

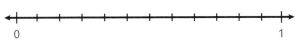

0   1

1. $\dfrac{2}{3} + \dfrac{1}{12}$

2. $\dfrac{1}{6} + \dfrac{1}{3}$

3. $\dfrac{3}{4} + \dfrac{1}{12}$

4. $\dfrac{5}{12} + \dfrac{1}{4}$

5. $\dfrac{1}{2} + \dfrac{1}{12}$

6. $\dfrac{1}{4} + \dfrac{2}{3}$

7. $\dfrac{5}{6} + \dfrac{1}{4}$

8. $1 - \dfrac{5}{6}$

9. $\dfrac{3}{4} - \dfrac{1}{3}$

10. $\dfrac{5}{6} - \dfrac{1}{2}$

11. $\dfrac{11}{12} - \dfrac{7}{12}$

12. $\dfrac{12}{12} - \dfrac{1}{3}$

13. $\dfrac{1}{4} - \dfrac{1}{6}$

14. $\dfrac{11}{12} - \dfrac{2}{3}$

Answer each question. Use a ruler if you wish.

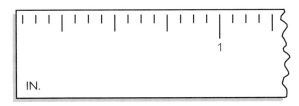

IN.

15. What is the sum of $\frac{5}{16}$ inch and $\frac{1}{4}$ inch?

16. What is the sum of $\frac{1}{2}$ inch and $\frac{3}{8}$ inch?

17. How much longer than $\frac{7}{8}$ inch is $\frac{3}{4}$ inch?

18. How much longer than $\frac{9}{16}$ inch is $\frac{7}{8}$ inch?

Extending Concepts

19. Find a path that has a sum equal to 2. You may enter any open gate at the top, but you must exit from the gate at the bottom.

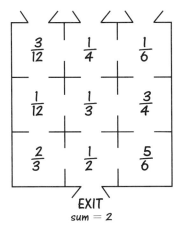

EXIT
sum = 2

Making Connections

20. Theo has a growth chart in the doorway of his closet. He likes to check his height every three months. In the year between his tenth and eleventh birthdays, he grew $\frac{1}{2}$ inch, $\frac{3}{4}$ inch, $\frac{3}{8}$ inch, and $\frac{5}{8}$ inch. How much did he grow that year?

21. In the thirteenth century, people used a unit of measure less than an inch. It was called a barleycorn because it was about the length of a barley seed. Three barleycorns make an inch.

a. What is the length in inches of 1 barleycorn? 5 barleycorns? 9 barleycorns?

b. Find the sum in inches of 1 barleycorn, 5 barleycorns, and 9 barleycorns.

Numbers Only

Applying Skills

Find each sum or difference.

1. $\dfrac{4}{9} + \dfrac{1}{3}$

2. $1 - \dfrac{3}{8}$

3. $\dfrac{5}{6} + \dfrac{5}{24}$

4. $\dfrac{8}{9} - \dfrac{1}{3}$

5. $\dfrac{5}{12} + \dfrac{1}{8}$

6. $1 - \dfrac{5}{6}$

7. $\dfrac{5}{6} - \dfrac{1}{3}$

8. $\dfrac{4}{5} + \dfrac{2}{3}$

9. $\dfrac{9}{40} - \dfrac{1}{10}$

10. $\dfrac{4}{5} - \dfrac{2}{15}$

11. $\dfrac{5}{12} + \dfrac{4}{8}$

12. $\dfrac{14}{15} - \dfrac{1}{6}$

13. $\dfrac{3}{7} + \dfrac{9}{14}$

14. $\dfrac{7}{10} - \dfrac{1}{6}$

15. $\dfrac{5}{8} - \dfrac{1}{2}$

16. $\dfrac{3}{7} + \dfrac{4}{5}$

17. $\dfrac{4}{5} - \dfrac{1}{6}$

18. $\dfrac{5}{8} - \dfrac{7}{12}$

19. $\dfrac{5}{9} + \dfrac{5}{6}$

20. $\dfrac{3}{5} + \dfrac{1}{15}$

21. Find $\dfrac{5}{8}$ minus $\dfrac{5}{12}$.

22. Find the sum of $\dfrac{9}{10}$ and $\dfrac{4}{15}$.

23. How much more is $\dfrac{3}{4}$ than $\dfrac{2}{3}$?

24. What is the sum of $\dfrac{3}{8}$ and $\dfrac{5}{6}$?

25. Evaluate $\dfrac{5}{8} - \dfrac{1}{4}$.

26. What is the sum of $\dfrac{2}{3}$, $\dfrac{5}{8}$, and $\dfrac{7}{12}$?

Extending Concepts

Find each missing fraction.

27. $\dfrac{4}{5} + \underline{\ \ ?\ \ } = 1$

28. $\underline{\ \ ?\ \ } + \dfrac{2}{3} = 1$

29. $\dfrac{2}{3} - \underline{\ \ ?\ \ } = \dfrac{4}{9}$

30. $\underline{\ \ ?\ \ } - \dfrac{3}{8} = \dfrac{3}{16}$

31. $\underline{\ \ ?\ \ } - \dfrac{3}{4} = \dfrac{1}{12}$

32. $\dfrac{1}{5} + \underline{\ \ ?\ \ } = \dfrac{13}{20}$

33. Find the path that has a sum equal to 1. You may enter any open gate at the top, but you must exit from the gate at the bottom.

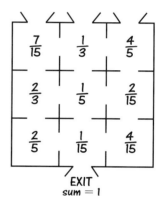

Making Connections

34. Benjamin Franklin liked to make up magic squares. His magic squares could be as large as 16 by 16. Here is a smaller one. Arrange the numbers,

$$\dfrac{1}{12}\ \dfrac{1}{6}\ \dfrac{1}{4}\ \dfrac{1}{3}\ \dfrac{5}{12}\ \dfrac{1}{2}\ \dfrac{7}{12}\ \dfrac{2}{3}\ \dfrac{3}{4}$$

into nine boxes so that the sum of each column, row, and diagonal is $1\dfrac{1}{4}$.

Not Proper but Still Okay

Applying Skills

Write each number in a different form: improper fraction, mixed number, or whole number.

1. $\dfrac{7}{3}$

2. $3\dfrac{5}{8}$

3. $\dfrac{55}{11}$

4. $5\dfrac{2}{5}$

5. $3\dfrac{7}{8}$

6. $\dfrac{24}{7}$

7. $\dfrac{35}{5}$

8. $11\dfrac{11}{12}$

9. $3\dfrac{9}{10}$

10. $\dfrac{44}{15}$

Find each sum. If the problem is written with improper fractions, write the sum as an improper fraction. If the problem is written with mixed numbers, write the sum as a mixed number. If the problem has both types of numbers, write the sum as an improper fraction and as a mixed number.

11. $2\dfrac{3}{4} + 3\dfrac{1}{2}$

12. $5\dfrac{1}{4} + 2\dfrac{5}{12}$

13. $2\dfrac{1}{2} + 5\dfrac{2}{3}$

14. $11\dfrac{3}{4} + 9\dfrac{3}{5}$

15. $\dfrac{5}{3} + \dfrac{11}{6}$

16. $\dfrac{22}{10} + \dfrac{57}{50}$

17. $\dfrac{7}{3} + \dfrac{12}{9}$

18. $\dfrac{7}{6} + \dfrac{9}{4}$

19. $4\dfrac{4}{5} + \dfrac{78}{15}$

20. $\dfrac{23}{7} + 3\dfrac{1}{4}$

21. $2\dfrac{3}{4} + \dfrac{19}{8}$

22. $\dfrac{7}{6} + 5\dfrac{5}{9}$

Extending Concepts

Find the path that has the given sum. You may enter any open gate at the top, but you must exit from the gate at the bottom.

23.

$\dfrac{1}{2}$	$2\dfrac{1}{4}$	$3\dfrac{1}{2}$
$1\dfrac{1}{3}$	$\dfrac{3}{4}$	$1\dfrac{5}{8}$
$5\dfrac{1}{8}$	$\dfrac{13}{24}$	$7\dfrac{2}{3}$

EXIT
sum = 10

24.

$\dfrac{5}{16}$	$2\dfrac{1}{8}$	$1\dfrac{3}{4}$
$3\dfrac{1}{4}$	$\dfrac{1}{16}$	$1\dfrac{3}{16}$
$\dfrac{2}{32}$	$\dfrac{1}{2}$	$2\dfrac{1}{32}$

EXIT
sum = 6

Writing

25. Answer the letter to Dr. Math.

> Dear Dr. Math,
>
> When I needed to solve the problem $2\dfrac{5}{16} + \dfrac{19}{4}$, a classmate said it would be easier if I changed $2\dfrac{5}{16}$ to an improper fraction. But, I do not want to do anything that is not proper! Do I have to solve the problem her way?
>
> Yours truly,
> Pollyanna DoRight

Sorting Out Subtraction

Applying Skills

Find each sum or difference. If a problem is written with mixed numbers, write the answer as a mixed number. If the problem is written with improper fractions, write the answer as an improper fraction. If the problem has both types of numbers, write the answer in either form.

1. $\dfrac{5}{4} - \dfrac{3}{8}$

2. $\dfrac{13}{5} - \dfrac{11}{10}$

3. $4 - 3\dfrac{5}{6}$

4. $8\dfrac{1}{3} - \dfrac{5}{9}$

5. $12\dfrac{3}{8} - 5\dfrac{3}{4}$

6. $\dfrac{33}{10} - 2\dfrac{3}{5}$

7. $\dfrac{5}{8} + \dfrac{3}{4}$

8. $2\dfrac{4}{5} - 1\dfrac{1}{6}$

9. $4\dfrac{9}{11} - \dfrac{35}{11}$

10. $1\dfrac{2}{3} + \dfrac{3}{4} + \dfrac{5}{4}$

11. $8\dfrac{8}{9} - 3\dfrac{1}{3}$

12. $2\dfrac{3}{13} + \dfrac{3}{2}$

13. $5\dfrac{1}{6} - 2\dfrac{1}{2}$

14. $7\dfrac{2}{5} - \dfrac{17}{10}$

15. $\dfrac{9}{4} + 5\dfrac{2}{3}$

16. $\dfrac{17}{6} - 1\dfrac{1}{9}$

17. $\dfrac{13}{4} - 2\dfrac{5}{6}$

18. $19 - \dfrac{12}{7}$

19. $4\dfrac{5}{9} + \dfrac{19}{6}$

20. $\dfrac{15}{4} + 19$

21. How much longer is $28\dfrac{1}{2}$ seconds than $23\dfrac{7}{10}$ seconds?

22. Find the sum of $4\dfrac{1}{5}$, $8\dfrac{7}{8}$, and $1\dfrac{7}{10}$.

23. What is the difference between $\dfrac{77}{9}$ and $5\dfrac{1}{3}$?

Extending Concepts

Complete each magic square. Remember that the sum of each column, row, and diagonal must be the same.

24.

$\dfrac{7}{2}$		$1\dfrac{1}{6}$
	$2\dfrac{11}{12}$	
		$\dfrac{7}{3}$

25.

5		$\dfrac{5}{3}$
$\dfrac{5}{6}$		
$6\dfrac{2}{3}$		

Making Connections

26. Kaylee is planning to bake a cake. She changed her recipe amounts to serve more people, and the resulting recipe involved many fractions. The first five ingredients are listed below.

Chocolate Cake

$2\dfrac{5}{8}$ cups flour

$1\dfrac{1}{8}$ cups cocoa powder

$1\dfrac{1}{2}$ cups sugar

$1\dfrac{1}{2}$ teaspoons baking soda

$\dfrac{3}{8}$ teaspoon salt

Kaylee has a bowl that holds 6 cups. Can she combine these ingredients in the bowl? Why or why not?

Calc and the Numbers

Applying Skills

Find each sum or difference. If a problem is written with mixed numbers, write the answer as a mixed number. If the problem is written with improper fractions, write the answer as an improper fraction. If the problem has both types of numbers, write the answer in either form.

1. $\dfrac{3}{5} + \dfrac{7}{8}$

2. $\dfrac{31}{32} - \dfrac{3}{16}$

3. $\dfrac{1}{30} + \dfrac{2}{30}$

4. $\dfrac{49}{50} - \dfrac{24}{25}$

5. $2\dfrac{3}{4} - 1\dfrac{1}{3}$

6. $\dfrac{8}{5} + 2\dfrac{2}{3}$

7. $\dfrac{15}{15} - \dfrac{3}{7}$

8. $1\dfrac{5}{8} + 3\dfrac{3}{4}$

9. $\dfrac{3}{7} + \dfrac{3}{4}$

10. $7\dfrac{1}{8} - 3\dfrac{3}{4}$

Extending Concepts

11. Three fourths of the students in Ms. Smith's class have standard backpacks. One eighth of the students have rolling backpacks. What fraction of the class has either a standard or a rolling backpack? What fraction of the class has neither?

12. Chapa received a check for her twelfth birthday. She spent $\dfrac{3}{8}$ of the money the day she got it. The following week, she spent $\dfrac{1}{4}$ of the total amount. What fraction of her money does she have left to spend?

13. The Hilliers are making pizza for dinner. Jill put anchovies on $\dfrac{1}{2}$ of the pizza. Then, Jack put sausage on $\dfrac{2}{3}$ of the pizza, including the half with the anchovies. What fraction of the pizza has sausage, but no anchovies?

14. At his graduation party, Gianni served $1\dfrac{5}{6}$ pounds of cheese and $2\dfrac{2}{3}$ pounds of peanuts. What is the total number of pounds of cheese and peanuts served?

15. Sophie has two watermelons. One weighs $3\dfrac{3}{4}$ pounds. The other weighs $5\dfrac{3}{8}$ pounds. What is the total weight of the watermelons? What is the difference between the weights of the two watermelons?

16. Analise spent $3\dfrac{1}{2}$ days visiting her grandmother. Then, she spent $\dfrac{5}{4}$ days at her uncle's house. How much time did she spend visiting these relatives? How much more time did she spend with her grandmother than her uncle?

Writing

17. Write a problem of your own, using a subject of your choice. Your problem should include at least two fractions, and its solution can be a fraction, mixed number, or a whole number.

Picturing Fraction Multiplication

Applying Skills

For items 1–2, use the drawing below.

1. How many cherries would represent $\frac{2}{3}$ of the cherries?

2. How many cherries would represent $\frac{5}{6}$ of the cherries?

For items 3–4, use the drawing below.

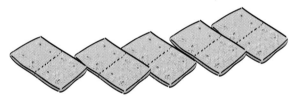

3. How many halves would represent $\frac{3}{10}$ of the crackers?

4. How many halves would represent $\frac{4}{5}$ of the crackers?

Solve each problem.

5. $\frac{4}{5}$ of 10

6. $\frac{3}{8}$ of 40

7. $\frac{2}{3}$ of 18

8. $\frac{5}{6}$ of 24

9. $\frac{3}{5}$ of 35

10. $\frac{3}{5}$ of 15

11. $\frac{1}{12}$ of 24

12. $\frac{3}{4}$ of 24

Extending Concepts

Solve each problem.

13. $\frac{2}{3}$ of 16

14. $\frac{3}{4}$ of 30

15. $\frac{3}{7}$ of 4

16. $\frac{7}{9}$ of 3

17. Danielle ate 11 of the orange wedges shown below. What fraction of the oranges did she eat?

18. Juanita served 3 giant submarine sandwiches at her party. Each sandwich was cut into 12 equal pieces. The guests ate $\frac{11}{12}$ of the sandwiches. How many pieces did they eat?

Making Connections

19. Myron manages a bakery. He has two pies that are exactly the same. However, the pies are cut into different size pieces and are priced differently. Which pie will earn more money?

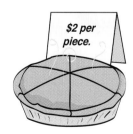

$2 per piece.

Three pieces for $4.

Fractions of Fractions

Homework 16

Applying Skills

For each problem, the rectangle represents one whole. Use the rectangle to find each solution.

1. $\frac{2}{3}$ of $\frac{3}{4}$

2. $\frac{3}{5}$ of $\frac{1}{6}$

3. $\frac{1}{9}$ of $\frac{2}{3}$

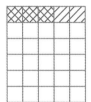

4. $\frac{5}{6}$ of $\frac{4}{5}$

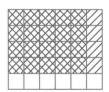

Find each product.

5. $\frac{3}{4} \times \frac{7}{8}$ **6.** $\frac{1}{2} \times \frac{4}{5}$

7. $\frac{2}{3} \times \frac{4}{7}$ **8.** $\frac{4}{9} \times \frac{3}{4}$

9. $\frac{2}{3} \times 60$ **10.** $\frac{3}{8} \times \frac{4}{9}$

11. $\frac{3}{5} \times \frac{10}{15}$ **12.** $\frac{1}{5} \times \frac{1}{5}$

13. $\frac{2}{3} \times \frac{1}{2}$ **14.** $\frac{1}{2} \times \frac{4}{9}$

Extending Concepts

15. Isabella spent $\frac{3}{4}$ of an hour doing homework. She spent $\frac{1}{2}$ of that time on math. What fraction of an hour did she spend on math?

16. Mia has a collection of baseball cards. Two-thirds of her collection represent National League players. One eighth of her National League cards represent the Giants. What fraction of her whole collection represents the Giants?

17. Ming has $\frac{1}{3}$ of a pizza left from last night's dinner. If he and his friend eat $\frac{1}{2}$ of the leftovers for an afternoon snack, how much of the whole pizza will they eat?

18. Jamil spends $\frac{1}{4}$ of his day at school. He spends $\frac{1}{10}$ of his school day at lunch. What fraction of his whole day does he spend at lunch?

Making Connections

19. About $\frac{3}{10}$ of Earth's surface is land.

a. At one time, $\frac{1}{2}$ of the land was forested. What fraction of Earth was forested?

b. Today about $\frac{1}{3}$ of Earth's land is forested. What fraction of Earth is forested today?

c. How much less of Earth is forested today?

Estimation and Mixed Numbers

Applying Skills

Choose the best estimate.

1. $\frac{15}{4} \times \frac{1}{2}$

   a. 2   **b.** 3   **c.** 8

2. $\frac{1}{3} \times \frac{5}{9}$

   a. 2   **b.** $\frac{1}{6}$ **c.** $\frac{1}{2}$

3. $2\frac{1}{4} \times \frac{1}{5}$

   a. $10\frac{1}{4}$ **b.** $\frac{1}{2}$ **c.** 1

4. $\frac{6}{5} \times \frac{3}{4}$

   a. $1\frac{1}{4}$ **b.** $\frac{18}{4}$ **c.** 1

Find each product.

5. $1\frac{1}{3} \times \frac{1}{2}$ **6.** $1\frac{4}{5} \times \frac{5}{9}$

7. $2\frac{3}{4} \times 3$ **8.** $1\frac{1}{5} \times \frac{1}{2}$

9. $\frac{2}{3} \times \frac{3}{4}$ **10.** $\frac{6}{7} \times \frac{9}{10}$

11. $\frac{3}{8} \times 2\frac{1}{4}$ **12.** $1\frac{1}{8} \times 1\frac{1}{8}$

13. $\frac{4}{3} \times 2\frac{1}{3}$ **14.** $\frac{3}{10} \times 20$

15. $3\frac{1}{4} \times 2\frac{2}{3}$ **16.** $2\frac{1}{2} \times \frac{5}{8}$

17. $5\frac{1}{3} \times \frac{4}{5}$ **18.** $2\frac{1}{2} \times 2\frac{2}{3}$

19. $3 \times 2\frac{1}{7}$ **20.** $3\frac{2}{3} \times 9$

Extending Concepts

The area of a rectangle equals the length times the width. Find the area of each rectangle.

21.

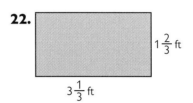

$1\frac{3}{4}$ in.

$2\frac{1}{2}$ in.

22.

$1\frac{2}{3}$ ft

$3\frac{1}{3}$ ft

Making Connections

23. Tia is making brownies for a crowd. She wants to triple her usual recipe shown below. Rewrite her recipe showing the amounts of each ingredient she will need.

Brownies
$\frac{2}{3}$ cup sifted flour
$\frac{1}{2}$ teaspoon baking powder
$\frac{1}{4}$ teaspoon salt
$\frac{1}{3}$ cup butter
2 squares bitter chocolate
1 cup sugar
2 well beaten eggs
$\frac{1}{2}$ cup chopped nuts
1 teaspoon vanilla

Fraction Groups within Fractions

Homework 18

Applying Skills

For items 1–3, use the following drawing to find each quotient.

1. $4 \div \frac{1}{4}$

2. $4 \div \frac{1}{2}$

3. $4 \div \frac{1}{8}$

For items 4–6 use the following drawing to find each quotient.

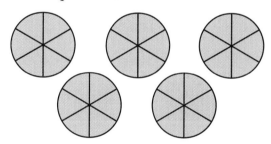

4. $5 \div \frac{1}{6}$

5. $5 \div \frac{1}{3}$

6. $5 \div \frac{1}{2}$

Find each quotient.

7. $\frac{2}{3} \div \frac{1}{3}$ **8.** $\frac{4}{3} \div \frac{1}{6}$

9. $2\frac{3}{4} \div \frac{1}{8}$ **10.** $\frac{4}{5} \div 3$

11. $\frac{5}{4} \div \frac{1}{3}$ **12.** $4\frac{1}{8} \div \frac{1}{4}$

13. $4\frac{1}{2} \div 3\frac{3}{4}$ **14.** $2\frac{4}{5} \div \frac{7}{8}$

Extending Concepts

Ms. Marrero is a math teacher who wants to celebrate pi day by buying pies for her class. (Pi day is March 14 or 3–14.) She has 27 students in the class. How much pie will each student get if she buys each of the following number of pies?

15. 3 pies   **16.** 5 pies

17. 9 pies   **18.** 30 pies

19. Maria has 7 pounds of mints. She wants to make $\frac{1}{4}$-pound packages for party favors. How many packages can she make?

20. Troy needs to cut a zucchini into slices that measure $\frac{3}{8}$ inch thick. If the zucchini is $6\frac{3}{4}$ inches long, how many slices will he have?

Writing

21. Imagine that your friend has been absent from school with a bad cold. You are taking the homework assignments to your friend's house. To do the math homework, your friend needs to understand how to divide fractions. To help your friend, draw a picture to show $4 \div \frac{1}{2} = 8$ and write a short note to explain the drawing.

Understanding Fraction Division

Applying Skills

For each problem, decide if the answer will be greater or less than the dividend. Then solve the problem to see if you are correct.

1. $\dfrac{5}{4} \div \dfrac{1}{2}$ **2.** $3\dfrac{2}{5} \div \dfrac{1}{5}$

3. $\dfrac{4}{5} \div 3$ **4.** $\dfrac{6}{5} \div 2\dfrac{1}{3}$

5. $\dfrac{9}{8} \div \dfrac{1}{4}$ **6.** $6 \div \dfrac{2}{3}$

7. $\dfrac{3}{4} \div 1\dfrac{1}{8}$ **8.** $\dfrac{2}{3} \div \dfrac{1}{6}$

9. $\dfrac{4}{3} \div \dfrac{1}{2}$ **10.** $3\dfrac{1}{2} \div 1\dfrac{1}{4}$

11. $\dfrac{1}{3} \div \dfrac{3}{5}$ **12.** $\dfrac{2}{5} \div 4$

13. $2 \div \dfrac{1}{6}$ **14.** $3 \div 4\dfrac{1}{2}$

15. $6\dfrac{1}{4} \div \dfrac{1}{2}$ **16.** $\dfrac{4}{5} \div \dfrac{6}{5}$

17. $4\dfrac{3}{4} \div \dfrac{5}{8}$ **18.** $5 \div \dfrac{2}{3}$

19. $5 \div 6\dfrac{1}{4}$ **20.** $1\dfrac{3}{8} \div \dfrac{3}{4}$

21. Is $\dfrac{7}{8} \div \dfrac{1}{2}$ greater or less than 1?

22. Is $\dfrac{3}{5} \div \dfrac{5}{6}$ greater or less than 1?

23. Find $\dfrac{3}{4} \div \dfrac{5}{6}$.

24. Find $3\dfrac{1}{4} \div 2\dfrac{1}{2}$.

Extending Concepts

To make two dozen muffins, you need $1\dfrac{1}{2}$ cups of flour. How many muffins can you make with each amount of flour?

25. 1 cup of flour

26. $\dfrac{1}{2}$ cup of flour

27. $\dfrac{1}{4}$ cup of flour

28. 2 cups of flour

29. $3\dfrac{1}{2}$ cups of flour

Writing

30. Answer the letter to Dr. Math.

> Dear Dr. Math,
> I have been doing math for a lot of years now, and I know a thing or two. One thing I know for sure is that when you do a division problem, the answer is always less than the first number in the problem. For example, $24 \div 4$ equals 6. Six is less than 24. That is how division works. So now, they are telling me that $10 \div \dfrac{1}{2}$ equals 20. I say no way, because 20 is greater than 10! I am right, right?
> Sincerely,
> M. Shure

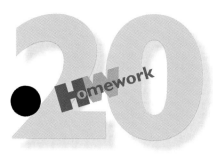

Multiplication vs. Division

Applying Skills

Find each value.

1. $\frac{1}{4} \times 4\frac{1}{4}$ **2.** $\frac{1}{5} \times \frac{2}{9}$

3. $\frac{4}{5} \div \frac{1}{10}$ **4.** $\frac{9}{8} \div \frac{1}{16}$

5. $\frac{4}{3} \times \frac{3}{4}$ **6.** $\frac{1}{2} \times \frac{1}{4} \div \frac{1}{8}$

For each pair of numbers, decide whether to multiply or divide to get the least number. Then, solve the problem. You must use the numbers in the order they are given.

7. $\frac{1}{4}$ and $\frac{1}{8}$ **8.** $\frac{2}{3}$ and $3\frac{1}{2}$

9. 5 and $\frac{1}{9}$ **10.** $\frac{2}{3}$ and 21

11. $\frac{5}{4}$ and $\frac{3}{8}$ **12.** $2\frac{1}{2}$ and $\frac{5}{2}$

Solve each problem.

13. Frank's bread recipe calls for $3\frac{1}{2}$ cups of flour and yields one loaf of bread.

a. How much flour will Frank need to make 7 loaves of bread?

b. How much flour will he need to make 13 loaves of bread?

c. Frank has $5\frac{1}{4}$ cups of flour. How many loaves of bread can he make?

d. How many loaves of bread can he make with 14 cups of flour?

14. There are 20 stamps on a sheet of stamps. Eli needs to mail 230 invitations. How many sheets of stamps will Eli use to mail the invitations?

15. Hernando has some cantaloupes that he plans to slice into 10 equal servings.

a. How many whole cantaloupes will he need to slice to have enough to give one slice to each of 27 people?

b. How much of a whole cantaloupe will be left?

16. About $\frac{7}{10}$ of the human body is water. If a person weighs 110 pounds, how many pounds are water?

Extending Concepts

For each set of numbers, write four true multiplication and division equations.

17. $\frac{1}{4}$, $1\frac{1}{2}$, and 6   **18.** $1\frac{1}{3}$, $1\frac{1}{2}$, and 2

Making Connections

19. The woodchuck is a large rodent native to North America. How much wood does a woodchuck chuck? Well, who knows? However, if a woodchuck *could* chuck wood, let's say it would chuck half a log of wood in 5 minutes. How much wood would a woodchuck chuck in $\frac{5}{12}$ of an hour?

Glencoe

The **McGraw·Hill** Companies

This unit of MathScape: Seeing and Thinking Mathematically was developed by the Seeing and Thinking Mathematically project (STM), based at Education Development Center, Inc. (EDC), a non-profit educational research and development organization in Newton, MA. The STM project was supported, in part, by the National Science Foundation Grant No. 9054677. Opinions expressed are those of the authors and not necessarily those of the Foundation.

CREDITS: Unless otherwise indicated below, all photography by Chris Conroy.

93 (b)Aaron Haupt; **94** Matt Meadows; **96** Getty Images; **105** ThinkStock LLC/Index Stock Imagery; **106** Aaron Haupt; **116** Images.com/CORBIS; **128** Matt Meadows.

Send all inquiries to:
Glencoe/McGraw-Hill
8787 Orion Place
Columbus, OH 43240-4027

ISBN: 0-07-866796-8

    5 6 7 8 9 10 058 09 08 07 06